Securing Our Water Supply

Securing Our Water Supply: Protecting a Vulnerable Resource

Dan J. Kroll

PennWell®

1421 South Sheridan Road
Tulsa, Oklahoma 74112-6600 USA
800.752.9764
+1.918.831.9421
sales@pennwell.com
www.pennwellbooks.com
www.pennwell.com

Director: Mary McGee
Managing Editor: Steve Hill
Production / Operations Manager: Traci Huntsman
Senior Design Editor / Book Designer: Robin Remaley
Production Editor: Tony Quinn
Cover Designer: Karla Pfeifer

Library of Congress Cataloging-in-Publication Data

Kroll, Dan J.
Securing our water supply: protecting a vulnerable resource / by Dan J. Kroll.
p. cm.
Includes bibliographical references and index.
ISBN-13: 978-1-59370-069-0 (hardcover)
ISBN-10: 1-59370-069-5 (hardcover)
1. Water-supply–Security measures–United States. 2. National security–United States. 3. Water-supply–Protection–United States. 4. Drinking water–Contamination–United States–Prevention. 5. Terrorism–Prevention. I. Title.
TD223.K766 2006
363.6'10973--dc22

2006013257

Printed in the United States of America

1 2 3 4 5 10 09 08 07 06

This book is dedicated to my daughter, Hannah, in the hope that when she grows up, worrying about things like terrorism and water quality will be a memory of the past, and also to my wife, Sandra, for putting up with me while I worked on this project.

Contents

Preface

Until September 11, 2001, the challenge of detecting intentional contamination of our water systems wasn't given much thought. For the past 17 years, as an employee of Hach Company, the world's largest manufacturer of analytical solutions for the drinking and wastewater industries, I have been involved with research directed at finding solutions to the problem of maintaining the quality of our nation's and the world's water supplies. My main role at Hach has been the development of analytical methods. In the past, most of these methods have focused on detecting and quantifying accidental and environmental pollutants in water and wastewater. For example, I developed the simple and inexpensive field test used by the Bangladesh Arsenic Mitigation Water Supply Project to test over five million wells in that country for naturally occurring arsenic contamination.

The tragic events in New York, Washington, and Pennsylvania rapidly and dramatically changed the picture for everyone in the water industry and made us aware of our vulnerability to deliberate contamination. At Hach, being experts in the water quality industry, we quickly recognized the vulnerability of our water supply systems to intentional contamination, and in January 2002, we began a program to develop strategies to combat this mode of attack. By April 2003, we formed a separate business unit, Hach Homeland Security Technologies, to specifically address these problems. I was named chief scientist of this group and have spent the past several years studying the terrorist threat to water and approaches to ameliorate it. The purpose of this volume is to share with you some of the insights I have gained into security issues as they relate to the water industry and hence, to better prepare for such an attack. As the proverb says, "Forewarned is forearmed."

I hope you find this book useful and informative.

Dan Kroll
Chief Scientist, Hach Homeland Security Technologies
DKroll@hach.com

1

The Psychology of Terrorism: Why Target Water?

While we must remain determined to defeat terrorism, it isn't only terrorism we are fighting. It's the beliefs that motivate terrorists. A new ideology of hatred and intolerance has arisen to challenge America and liberal democracy.

—John Kerry

Introduction

The events of September 11, 2001, dramatically and irrevocably changed our worldview. The picture that many of us had of an America safe and secure from foreign attack was shattered on that morning. We realized, like never before, that we are vulnerable. As the days following the attack progressed, we came to understand the extent and the scope to which we were open to devastating terrorist assaults were not limited to kamikaze jetliners crashing into buildings. As an open society, the critical infrastructure we need for our culture and way of life to endure was not hardened. Our power grid, cyberspace, the transportation network, communications systems, public health, the food supply, and the water supply (the focus of this book) were open to assaults that could rapidly cripple the nation. The country quickly mobilized to address these issues.

Since then, we have made great strides in securing certain weak points, but much work remains to be done. Today we find ourselves in the midst of what President Bush has termed "The War on Terror." Our troops fight this war on the battlefields of Afghanistan and Iraq and in myriad other places around the globe. Those of us at home do our part by trying to decrease our vulnerability to future attacks and regain some vestige of the sense of security that we lost on 9/11.

As a starting point in our fight to regain security, it is imperative to understand the threat that terrorism poses. To get a clear picture of the threat, it is important to have a historical perspective on what terrorism is and how it has changed over the centuries. It is also helpful to understand the psychology that motivates terrorists to take actions that can lead to such devastation. This is not an easy task. In the words of the great 19th-century Russian author Fyodor Dostoevsky, "While nothing is easier than to denounce the evildoer, nothing is more difficult than to understand him." Such an undertaking was presaged and deemed to be imperative by Sun Tzu, the Chinese sixth-century-BC author of *The Art of War,* who wrote, "If you know the enemy and know yourself you need not fear the results of a hundred battles."

A Historical Perspective on Terrorism

Terrorism is not easily defined. Most people rely on the approach to defining terrorism that Supreme Court Justice Potter Stewart used when defining pornography—that is, "I'll know it when I see it." While there have been many attempts at defining terrorism, the U.S. government's official definition is as good as any. The U.S. Code of Federal Regulations contains the following specific definition of terrorism: "the unlawful use of force and violence against persons or property to intimidate or coerce a government, the civilian population, or any segment thereof, in furtherance of political or social objectives."[1]

With this definition in mind, it becomes obvious that terrorism is not a recent phenomenon. It has been with us in some form or another since the first leaders of primitive societies reinforced their rule with examples of violence or brutality. Xenophon (an ancient Greek historian [431–350 BC]) wrote of the effectiveness of fear and psychological warfare in subduing enemy populations.[2] The rulers of many ancient societies used the tactics of terrorism to reinforce their control of the general populace. The people, in turn, used these same tactics in resisting their rulers. The complete destruction of Carthage by the Romans in 146 BC is an early example of these tactics taken to the extreme. Rome was a sworn enemy of the Carthaginian Empire, which existed across the Mediterranean in what is presently Tunisia. Rome and Carthage fought a series of wars known as the Punic wars. During the third and final Punic War, after the sacking of Carthage, the Roman army slaughtered the inhabitants, burned the city to the ground, and sowed the surrounding countryside with salt so that nothing would grow on the site (fig. 1–1).[3] This was to dissuade others from challenging Rome.

Throughout the history of their empire, the Romans used terrorism to subjugate and intimidate the populace of their conquered territories. The Roman emperors Tiberius (AD 14–37) and Caligula (AD 37–41) were notable in their use of terrorist tactics such as summary public executions, banishment, and confiscation of property. Their motivation was to cow their subjects into submission and to prevent the fulmination of rebellion in the conquered territories.[2]

Fig. 1–1. Roman legions destroyed the city of Carthage and the Carthaginian culture in 146 BC, at the end of the third Punic War. (Image from www.nationmaster.com)

However, it wasn't long before the Romans themselves became the victims of terrorist tactics. In the first century AD, the Sicari and the Zealots, Jewish resistance groups active in and around Jerusalem, employed terrorist tactics to disrupt Roman rule. Their tactic of choice was public assassination. The Sicari targeted Jews they considered to be collaborators, and the Zealots targeted Romans and Greeks. The assassinations were carried out in broad daylight in front of witnesses to make sure that a message was sent.[4] Many of the earliest examples of terrorism, like the Sicari and the Zealots, had their basis in religion. The Roman persecution of the early Christian Church is a prime example.

Religion was also the motivating factor behind a Hindu sect that was active from the 7th century to the 19th century. This sect, known as the Thugees (hence the word thug), practiced ritualized murders in their worship of the Hindu goddess of destruction and terror, Kali (fig. 1–2). The victims of these ritual murders were required to be subjected to an absolute state of terror before they were killed. There is some debate as to whether the Thugees represent an actual terrorist group, because they had no ulterior motive in their use of terror other than the practice of their religion. While the victims were commonly robbed before being killed, this was not for personal gain but as an offering to Kali. The modes of killing were ritually prescribed, and who could be a victim was also dictated. There was a strict rule that foreigners were not worthy sacrifices and could not be killed. This led to the sect's downfall when India was under British rule, because British officers could track and arrest the members without being afraid that they would become victims.[4]

Fig 1–2. Kali, Hindu goddess of destruction and terror

Another example of terrorism motivated by religious principles occurred in the 11th century. At this time, a Shia Muslim sect, the Ismailis, spawned a violent offshoot known as the "Hashish Eaters" or, in the local dialect, "Assassins" (giving us the modern word assassin). The Assassins were sworn to the forcible spread of their version of Islam. Tactics included public killings, often perpetrated at religious sites on holy days to garner large audiences so that their message would be widely disseminated in the hope that others would follow their example.[4]

Secular terrorism (not motivated by religion), along with the use of the word "terror" in reference to a systematic policy, has its roots in the French Revolution and the ensuing Reign of Terror. Maximilien Robespierre (fig. 1–3) made a famous justification for the use of terror in a speech he gave in front of the French National Convention in 1794. According to Robespierre,

> If the basis of a popular government in peacetime is virtue, its basis in a time of revolution is virtue and terror—virtue, without which terror would be barbaric; and terror, without which virtue would be impotent.[5]

The word terror is based on the Old French (14th century) word *terreur*, meaning great fear.[6] In general, throughout the centuries, terrorism has been used by governments, revolutionary movements, small groups, and individuals to realize avowed political or social goals that were deemed not achievable through political or conventional military means.

Fig. 1–3. Portrait of Maximilien Robespierre by Adelaide Labille-Guiard (1786). Robespierre was an advocate of the use of secular terrorism.

Nationalist movements, labor unrest, and the birth of Marxism is the 1800s gave rise to numerous terrorist organizations. The Molly McGuires in the coalfields of the eastern United States, the Russian anarchists, and the Feinian Movement in Ireland all flourished at this time. It was also at this time that Carlo Pisacane, an Italian independence fighter, espoused theories that are the basis of modern terrorism. He was the founder of the concept of *propaganda by deed*. His fundamental tenet was that ideas by themselves are not adequate to move a revolutionary ideal forward.

They are spread in too weak a form and reach too few people to lead to effective action, much less to results. It is only by action that the masses can be swayed.[7] These ideas led to the growing importance of media coverage in dictating what form terrorist actions would take.

Terrorism is basically the use of violence against noncombatants in such a manner as to instill fear and force the desired conclusion in a political confrontation. The state practice of coercion of the populace by terrorist means (as in ancient Rome, Nazi Germany, Stalinist Russia, Mao's China, Pol Pot's Khmer Rouge in Cambodia, and Saddam's Iraq) and the use of terrorist tactics by nationalistic resistance movements (Zealots, Russian Anarchists, Irish Republican Army [IRA]) have been constant in world history, as has the targeting of specific racial or religious groups by terrorist organizations (e.g., the formation of the Ku Klux Klan [KKK] in the American South after the Civil War [fig. 1–4]). Terrorist groups based solely on religious motivations, while common in early terror history, tended to die out in modern times; however, they have recently reemerged with a vengeance.

*Fig. 1–4. Mississippi KKK in the disguises in which they were captured (*Source: Harper's Weekly, *January 27, 1872)*

While many past terrorist organizations were in some way based on religion (the Zealots were a Jewish group, the KKK is predominately Protestant, and the IRA is predominately Catholic), the use of terror based solely on religion virtually disappeared from the scene with the demise of the Thugees in India in the 1800s.[4] Its reemergence in the late 20th century is troubling. This rise of terrorist organizations with religion as their primary motivation has resulted in a morphing of terrorist goals and modes of operation. Past terrorist organizations worked on the operational basis of the Chinese proverb that states that the goal is "to kill one and frighten 10,000." Mass casualties were not a goal of past terrorist organizations.

Terrorism and Weapons of Mass Destruction

Popular dogma in the 1970s and '80s held that terrorists would forgo the utilization of weapons of mass destruction (WMDs), as their use would, in the long run, prove to be a tactic that could backfire. According to Brian Jenkins, one of the world's highest authorities on political violence and sophisticated crime and an advisor to the National Commission on Terrorism, "Terrorists want a lot of people watching, not a lot of people dead."[8] The use of such weapons by the nationalistic and separatist movements that dominated the world terror scene at the time would have been counterproductive. Such tactics would have alienated the popular sentiment at home that these groups require to survive. Infliction of mass casualties would most likely have caused international sources of funding that support the activities of groups like the IRA and others to dry up as well. These were chances that terrorist organizations of that period were unwilling to take.

This self-imposed moratorium on mass murder doesn't hold for newer organizations grounded in religion. The rebirth of religious terrorism, with different goals from the traditional terrorist groups, has led to an escalation in weapons use and tactics to the point where there are grave fears of a terrorist organization resorting to the use of a nuclear weapon or biological agent that could cause casualties in the millions. What has changed in the terrorist motivation and psyche that has caused this escalation?

Terrorist organizations based on religion, whether they are cults or fundamentalist in nature, have goals and aspirations that are, for the most part, not as limited as those of secular groups. According to Rex Hudson, in the report *The Sociology and Psychology of Terrorism*,

> When the conventional terrorist groups and individuals of the early 1970's are compared with terrorist of the early 1990's, a trend can be seen: the emergence of religious fundamentalist and new religious groups espousing the rhetoric of mass-destruction terrorism.
> In the 1990's groups motivated by religious imperatives, such as

> Aum Shinrikyo, Hezbollah, and al Qaeda, have grown and proliferated. These groups have a different attitude towards violence—one that is extranormative and seeks to maximize violence against the perceived enemy, essentially anyone who is not a fundamentalist Muslim or an Aum Shinrikyo member. Their outlook is one that divides the world simplistically into "them" and "us". With its sarin attack on the Tokyo subway system on March 20, 1995, the doomsday cult Aum Shinrikyo turned many expert's prediction of a religion based terrorist organization using WMD into a reality.[9]

The goals of traditional terrorist groups are usually moderate in scope. These organizations typically seek independence from an occupying power, the redress of perceived social wrongs, labor rights, and so forth. A generation ago, terrorists' wishes were very limited, specific, and clear. On hijacking three airliners in September 1970, for example, the Popular Front for the Liberation of Palestine demanded, with success, the release of Arab terrorists imprisoned in Great Britain, Switzerland, and West Germany. On attacking the B'nai B'rith headquarters and two other Washington, D.C., buildings in 1977, a Hanafi Muslim group demanded the canceling of a feature movie, *Mohammad, Messenger of G-d,* $750 (reimbursement for a fine), the turning over of the five men who had massacred the Hanafi leader's family, and the killer of Malcolm X.[10]

The stated goals of al Qaeda are an example of how expansive some of the organizations that are based on religion have become in their philosophy. Al Qaeda has stated that its immediate goals are to remove infidels from the holy land and establish Islamic rule throughout the Middle East. This includes ensuring a homeland for the Palestinians. These goals, while ambitious, are not dramatically different from goals of traditional nationalist terrorist organizations. Where they diverge from normal is in their long-term aspirations. Osama bin Laden's long-term goals include the complete destruction of Western Civilization and the establishment of a Worldwide Islamic Caliphate that would rule on Islamic principles with roots in the 14th Century. Those countries not under direct Muslim rule would be subservient to the caliphate. A religious scholar that is greatly admired and often quoted by bin Laden is Sheikh Abdullah bin Muhammad bin Humaid. In an essay entitled "The Call to Jihad," Humaid argues that all Muslims are obligated to participate in a perpetual *jihad* against the non-Muslim world. As he explains,

> Allah revealed in Surat At-Taubah the order to discard all the obligations and commanded the Muslims to fight against all the Mushrikun as well as against the people of the Scriptures (Jews and Christians) if they do not embrace Islam, till they pay the Jizyah (a tax levied on the non-Muslims who do not embrace Islam and are under the protection of an Islamic government) with willing submission and feel themselves subdued.[11]

In nearly all cases, jihadi terrorists have a patently self-evident ambition: to establish a world dominated by Muslims, Islam, and the Shari'a (Islamic law). Or, in the words of the *Daily Telegraph,* their "real project is the extension of the Islamic territory across the globe, and the establishment of a worldwide 'caliphate' founded on Shari'a law." Terrorists openly declare this goal. The Islamists who assassinated Anwar el-Sadat in 1981 decorated their holding cages with banners proclaiming "the caliphate or death." A biography of Abdullah Azzam, one of the most influential Islamist thinkers of recent times and an influence to Osama bin Laden, declares that his life "revolved around a single goal, namely the establishment of Allah's Rule on earth" and restoring the caliphate. Bin Laden himself spoke of ensuring that "the pious Caliphate will start from Afghanistan." His chief deputy, Ayman al-Zawahiri, also dreamed of reestablishing the caliphate, for he wrote, "history would make a new turn, God willing, in the opposite direction against the empire of the United States and the world's Jewish government." Another al Qaeda leader, Fazlur Rehman Khalil, has published a magazine that declares, "Due to the blessings of jihad, America's countdown has begun. It will declare defeat soon," to be followed by the creation of a caliphate.[10]

In the quest to realize these inflated ambitions, Osama bin Laden is by no means deterred by the potential of mass casualties and has been widely known to espouse the use of WMD. In fact, mass casualties are a stated goal of al Qaeda. Suleiman Abu Ghaith, one of Osama bin Laden's closest friends and allies, said on an Islamic Web site that the terrorists planned to attack the United States with chemical or biological weapons. In his words, 9/11 was

> only a start. We have the right to kill 4 million Americans—2 million of them children—and to exile twice as many and wound and cripple hundreds of thousands. It is our right to fight them with chemical and biological weapons, so as to afflict them with the fatal maladies that have afflicted the Muslims because of the Americans' chemical and biological weapons.[12]

Statements such as these make it obvious that in this struggle no holds are barred, and we can expect these groups to use WMDs or other means of effecting mass casualties wherever and whenever they can.

Osama bin Laden actually sought and received approval for the use of WMD from Islamic Clerics. An Islamic religious ruling granted Osama bin Laden and other terrorist leaders permission to use nuclear, biological, or chemical weapons against the United States and its allies. The existence of this ruling, or *fatwa,* was revealed by Michael Scheuer, a former Central Intelligence Agency (CIA) officer who headed the agency's bin Laden unit from 1996 to 1999, during a November 14, 2004, broadcast of the CBS news show *Sixty Minutes*. The fatwa was issued by a prominent Saudi cleric, Sheikh Nasir bin Hamid al Fahd, in 2003 and is an ominous sign that bin Laden and al Qaeda no longer accept moral or religious values obstructing the use of WMD. Al Fahd is part of an ultraconservative trio of Saudi religious leaders (al Fahd, Ali al-Khudayr, and Ahmadal Khaladi), in what is known as the *Salafi Jahadi* trend. This movement is part of the recent emergence of a self-proclaimed "utterly pure Islamic model" that began to appear in the late 1990s, at the same time the Taliban regime emerged in Afghanistan. The Salafi school was formed to reply "to any criticism of the Taliban's measures that did not enjoy the approval of prominent scholars of the Islamic world. The movement simply served to justify any actions condemned by mainstream Islam. To them, the Taliban regime was the most legitimate representative of Islam and anyone who joins the cause of fighting international terrorism, that is, dismantling al Qaeda and the Taliban, is an *infidel.* In the mind of Osama bin Laden and to al Qaeda, the religious permission of the fatwa removed any last impediment to using WMD.[13]

The Terrorist: Who and Why?

It is very difficult to answer the question of who becomes a terrorist and why. Most people do, however, have a picture in their mind's eye that they use to envision a typical terrorist. The general public's broad view of a terrorist is a deranged fanatic; if they are not mentally insane, then they are pretty darn close. Terrorists must be impoverished, unsophisticated, uneducated people who have been led astray due to their ignorance. Is this image truly representative of the actual terrorist we are facing?

There have been psychological studies into what type of person becomes a terrorist. As to the mental condition of terrorists, some experts, such as Gerald Post director of the political psychology program at George Washington University, suggest that to commit such acts, the terrorist must suffer some form of mental disorder. Post states, "political terrorists are driven to commit acts of violence as a consequence of psychological forces, and that their special psycho-logic is constructed to rationalize acts they are psychologically compelled to commit." He feels that the most virulent forms of terrorism are generational in nature. Hatred is passed on from generation to generation.[14] While this is the view of many psychologists and mirrors the view held by the public, just as many psychologists disagree.

Many experts hold that as a group, terrorists are generally quite sane. Martha Crenshaw, international terrorism expert and professor of government at Wesleyan University states, "The outstanding common characteristic of terrorists is their normality."[15] Maxwell Taylor (head of the Applied Psychology Department at University College, Cork, and Ethel Quayle agree: "The active terrorist is not discernibly different in psychological terms from the non-terrorist. In psychological terms, there are no special qualities that characterize the terrorist."[16] The general consensus is that while mentally ill individuals can commit terroristlike actions, they are unlikely to belong to organized terrorist groups. The selection process for these organizations would tend to weed out the mentally unfit. The group refuses to jeopardize itself or its goals by allowing in such people. Just because terrorists appear to be alienated from society as a whole doesn't necessarily translate into a mental illness.[9]

As far as socioeconomic background and education are concerned, several studies have not supported the popular stereotype of the terrorist as poor and uneducated. Terrorists as a group tend to have a greater than average education. A study conducted by Russell and Miller, called *Profile of a Terrorist,* found that around two-thirds of the terrorists studied had at least some college education.[17] Also, there seems to be little correlation between what sort of job a person holds and their proclivity to commit terrorist acts, although the unemployed and students tend to be recruited in higher numbers.

Russell and Miller also found that the majority (more than two-thirds) of the terrorists studied did not come from deprived economic groups. In fact, they were mostly from middle-class or even upper-class socioeconomic backgrounds. This is evidenced by those implicated in the London subway and bus bombings of 2005. These men were not economically deprived but were of middle- and upper-middle-class backgrounds. As Nasra Hassan, who interviewed terrorists and their families, noted in *The Times,*

> None of the suicide bombers—they ranged in age from 18 to 38—conformed to the typical profile of the suicidal personality. None of them was uneducated, desperately poor, simple-minded, or depressed. Many were middle-class and held paying jobs. Two were the sons of millionaires. They all seemed entirely normal members of their families. They were polite and serious, and in their communities were considered to be model youths. Most were bearded. All were deeply religious.[18]

There are some exceptions to this, however. Very large nationalist revolutionary movements, such as the Fuerzas Armadas Revolucionarias de Columbia (FARC [Revolutionary Armed Forces of Columbia]) and also the rank and file of Islamic fundamentalist organizations, tend to be of a more impoverished nature. This is especially true of Arab terrorist organizations that contain large numbers of the poor and unemployed.[17] It is important to note, however, that the leaders and operational planners of these organizations are almost all from the upper and middle classes.

In general, there is no standard for the level of education or economic background of a terrorist. Profiling can be a futile effort as the diversity represented by the population of terrorists is so large. It is important to be aware of this diversity, so as not to underestimate the capabilities of terrorists. For the most part, they are not ignorant and are very skilled in the art and craft of terrorist activities. Even those groups that are predominately made up of operatives with a lower skill level are aptly led and organized. It would be a grave error to let preconceived notions cloud our views as to terrorists' capabilities.

Americans have a long history of underestimating enemies. When the Japanese attacked Pearl Harbor, we were incredulous. According to Winston Groom's book *1942: The Year That Tried Men's Souls,* after Japan's successful opening attacks on American forces in the Pacific, both General Douglas MacArthur and Winston Churchill believed that Germans, not Japanese, had piloted the Japanese planes. Since the American media, both Hollywood and the press, had long portrayed the Japanese as hapless, many thought they were incapable of sophisticated military attacks. Groom writes, "The typical Japanese soldier was depicted in newspaper cartoons as a short, bandy-legged, buck-toothed, nearsighted, chattering ape." The supposedly inept Japanese not only flew the planes but also had designed and manufactured aircraft that were superior to anything in the U.S. arsenal at the time.[19] If we would have had more respect for our enemies and a better understanding of who they were and their capabilities, some of the early mistakes made in World War II could have been avoided.

It behooves us not to make similar errors in our assessment of the capabilities of modern terrorists. Underestimation of the enemy is a pitfall that is especially dangerous, because it automatically leads to a sense of security. This false sense of security can result in mistakes that could prevent us from taking appropriate measures to prevent attack.

The discussion so far has been focused on the international terrorist, but they are by no means the only danger. There is a wide diversity and ever-growing number of homegrown terrorist threats. Members of Far Right and Far Left organizations, ecoterrorists, narcoterrorists, gang members, deranged loners, and disgruntled workers may also pose a significant risk. In fact, according the Federal Bureau of Investigations (FBI), going by the number of incidents alone, extremist groups, such as the Animal Liberation Front (ALF) and the Earth Liberation Front (ELF), represent the top domestic terrorist threat.[20] While these groups have as yet avoided actions that would cause the loss of human life, they are beginning to get more violent, using tactics such as arson and bombing attacks, and have been known to cause large-scale property damage. An attack on the water supply would not be outside their modes of operation.

International terrorists and fringe groups are not the only sabotage concerns in the water industry. A possible attack orchestrated by an insider is also a vulnerability that needs to be addressed. An insider, such as a disgruntled worker, would be familiar with water supply systems and would know the vulnerable points for attack. While the danger of attempts by such individuals to cause mass casualties is slight (most past cases of workplace violence leading to deaths have been linked to people with diagnosable mental illness, which seems to be a requirement to take such extreme measures in their own community), they do represent a distinct possibility as far as nuisance attacks and denial of service actions are concerned. The question remains, however: regardless of who the terrorists are or what their goals are, is water a viable terrorist target?

Terrorism as Theater: Does Water Qualify?

Pisacane's theory of propaganda by deed began the long-standing courtship of media attention by the perpetrators of terrorist acts. Terrorists have always sought to maximize the exposure of their deeds. That is why many historical acts were orchestrated to achieve the largest audience possible—for example, the Assassins' choice of the optimum venue to publicize their attacks, committing their murders at shrines on holy days. The rise of the mass media has allowed terrorists to play to a truly worldwide audience. The importance of an audience to the terrorist agenda was put into perspective when Abbe Hoffman made the statement that all terrorist attacks are at least partly theater.[21]

A terrorist attack is intended to serve more than the single purpose of simply scaring the attacked individuals. It is meant to have meaning for the victim's peer group as well so that not just the victim, but also all those whom the victims are representative of are subjected to the terror of the deed and are plunged into a state of fear. It is also meant to have meaning to observers who are actually or potentially sympathetic with the terrorists' cause, thus helping in the recruitment of new terrorists. The result must be easily communicated to and understood by both groups. Therefore, most attacks are chosen because they are easily interpreted by the intended audience. The visual media is the most universal for conveying a message to a wide audience. This leads to the choice of tactics and targets that are easily conveyed in video and still pictures. A crashed airliner, a burning building, or explosions do not require explanation in multiple languages to be understood by an international audience. For example, the current insurgency in Iraq has been careful to make detailed video recordings of brutal and graphic terrorist acts such as the beheading of captives. These recordings then appear in videos that are used as training and recruiting tools.[22]

Many people hold that it is just for this reason that water is not a likely terrorist target. The argument is that an attack on water supplies does not fill the bill as an example of good theater. It lacks the dramatic special effects of a fiery airplane crash or the human drama of a hostage crisis. Their argument further states that the only way to enhance the theatrical effect of an attack on water supplies is to orchestrate it in such a way as to have it coincide with another type of attack. For example, pipelines could be blown up to disrupt the water supply or be contaminated to the extent that the system's operators are forced to shut it down. In close coordination with the denial of service attack on the water system, mass bombings or arson could be carried out in the area affected by the loss of water. The lack of water combined with the more conventional bombing or arson attacks would prevent the timely dousing of the flames and add to the general destruction. The idea is that this would generate more terror than a contaminant attack on water supplies initiated with the sole goal of causing mass casualties.[23]

On this point I disagree. An attack orchestrated so as to contaminate a water supply to the extent that it were to cause mass casualties does make good theater. Our culture has been inured to overt acts of violence by television and movie's graphic depictions of such events. Fiery explosions, even on the scale of those of 9/11, affect us emotionally not because of their Spielbergesque graphic nature but because we associate the pictures with the thousands of lives we know were lost in the event. A water event that killed thousands or tens of thousands would no doubt affect us in the same way. While photographs of casualties lined up on stretchers, long lines waiting for triage at hospitals, doctors working fervently to save lives, mass funerals, the emotion of empty classrooms and the grief of survivors wouldn't have the blockbuster special effects impact of the twin towers crashing down, it would still be highly dramatic and would rivet worldwide interest for some time.

Perhaps even more important is that people are amazingly resilient if given a chance. Most big-impact events, even those as dramatic as 9/11, can be intellectualized into triviality by the majority of the population. After 9/11, many people thought, "My God, this is terrible. Thank goodness it can't happen to me. I seldom if ever fly, and I don't live in a big city or work in a tall building. I am personally safe." People in small-town and rural America were, without a doubt, affected by the 9/11 attacks, but the feeling of personal danger was transient for most people. The intellectualization process quickly restored people to the point where they could proceed with their everyday lives.

For the vast majority of Americans, the same rationalization process wouldn't work for a mass-casualty water event. The process of distancing yourself from a danger posed by drinking water is not an easy matter. Everyone drinks water and/or consumes products that contain water. According to the United States Geological Survey, in the year 2000, about 84% of the population (or 240 million people) relied on public water supplies for their needs (fig. 1–5).[23] Even those who have a private source of water, such as a well supplying a single house, still drink water from other sources and consume commercial products containing water.

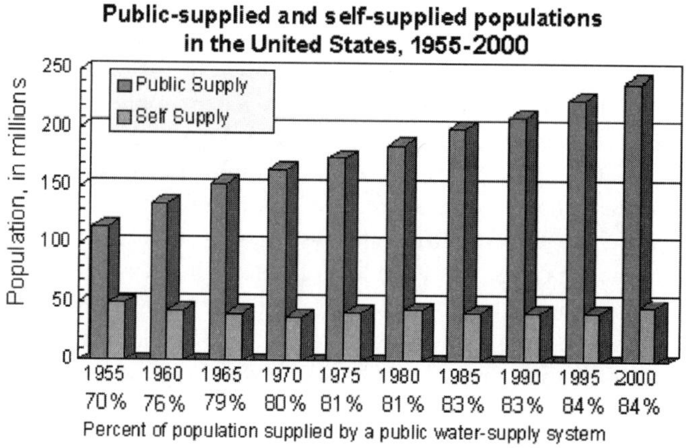

Fig 1–5. Graph of public water usage and needs (Source: *United States Geological Survey)*

If a mass casualty water event were to occur anywhere in the United States, the effects to the psyche of the nation could be devastating. Few people, if any, could completely rationalize away the feeling of personal endangerment that would result from such an event. The result of this could be psychological effects that would be more widespread and of a longer duration than even those that arose after the 9/11 attacks.

In their investigation of the attacks, the 9/11 Commission concluded that al Qaeda has specific attack criteria that they evaluate when planning an attack. These six criteria are used by al Qaeda planners to evaluate whether a mission will be effective:[24]

1. The attack cannot be trivialized or marginalized
2. The attack must be capable of international media understanding
3. The attack must be relatively inexpensive
4. The attack must be technologically simple
5. The attack ideally will exploit (and undermine) the attributes of a capitalist society and the West
6. The attack must incorporate multiple targets simultaneously

Hopefully after reading this book you will come to the same conclusion that I have, that on the basis of these criteria, water emerges as an ideal target for an al Qaeda–type attack. As the document shown in figure 1–6 indicates, the terrorists have shown an interest in attacking water supplies.

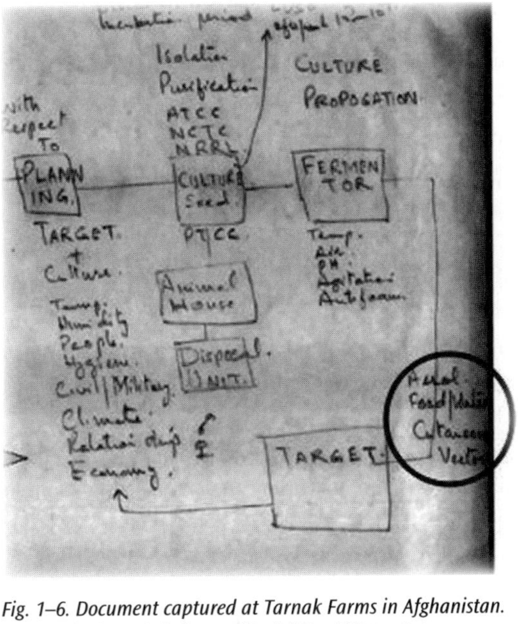

Fig. 1–6. Document captured at Tarnak Farms in Afghanistan. The text in the red circle says "Aerial Food/Water Cutaneous Vector," clearly indicating water as a target.

The recognition that our water supplies are vulnerable to sabotage is not simply a recent discovery in light of the attacks of 9/11. As early as 1941, after America had suffered another devastating surprise attack, FBI Director J. Edgar Hoover wrote,

> Among public utilities, water supply facilities offer a particularly vulnerable point of attack to the foreign agent, due to the strategic position they occupy in keeping the wheels of industry turning and in preserving the health and morale of the American populace. Obviously, it is essential that our water supplies be afforded the utmost protection.[25]

The U.S. military also recognized the threat prior to 9/11. The Department of Defense's historic policy requiring domestic military bases to rely on local utility infrastructure whenever possible for water supplies and other needs was explored and recognized as a threat by Major Donald C. Hickman, in his seminal report.[26] In fact, deliberate attacks on water are not just conjecture; they have occurred in the past and continue to occur today.

Notes

1. Code of Federal Regulations 28, section 0.85.
2. Encyclopaedia Britannica Online Edition. Terrorism. http://www.britannica.com/eb/article?tocId=217764
3. Illustrated History of the Roman Empire. http://www.roman-empire.net/index.html
4. Center for Defense Information. 2003. Terrorism: A brief history of terrorism. July 2, 2003. http://www.cdi.org/friendlyversion/printversion.cfm?documentID=1381
5. Robespierre, Maximilien. 1794. Speech in the French National Convention.
6. The Online Etymology Dictionary. http://www.etymonline.com/
7. Hoffman, Bruce. 1988. *Inside Terrorism.* New York: Columbia University Press.
8. Jenkins, Brian M. 1975. *High Technology Terrorism and Surrogate Warfare: The Impact of New Technology on Low-Level Violence.* Santa Monica, CA: Rand.
9. Hudson, Rex. 1999. *The Sociology and Psychology of Terrorism: Who Becomes a Terrorist and Why?* Report prepared under an interagency agreement by the Federal Research Division, Library of Congress.
10. Pipes, Daniel. 2005. The attempt to establish a world dominated by Muslims, Islam, and the Shari'a has begun—but the world is in denial. *Jewish World Review.* July 26.
11. Gartenstein-Ross, Daveed. 2004. Osama's big lie. *FrontPagemagazine.com.* December 17. http://www.frontpagemag.com/Articles/ReadArticle.asp?ID=16356

12. Lines, A. 2002. Al Qaeda: We'll kill 4 million more Americans" *Daily Mirror*. June 14. Hagman, Daniel. 2005. HQ intelligence alert. *Weekly Intelligence Briefing*. 2, issue 30 (August).

13. Post, Jerald. 1990. Current understanding of terrorist motivation and psychology: Implications for a differentiated antiterrorist policy. *Terrorism*. 13 (1): 65–71.

14. Crenshaw, Martha. 1981. Causes of terrorism. *Comparative Politics*. July 13.

15. Taylor, Maxwell, and Ethyl Quayle. 1994. *Terrorist Lives*. London and Washington: Brassey's. London.

16. Russell, Charles A., and Bowman H. Miller. 1977. Profile of a Terrorist. *Terrorism: An International Journal*. 1 (1): 17–34.

17. Hassan, Nasara. 2005. Are you read? Tomorrow you will be in paradise. *London Times*. July 16.

18. Groom, Winston. 2005. *1942: The Year That Tried Men's Souls*. New York: Atlantic Monthly Press.

19. Associated Press. 2005. Activist extremists top U.S. domestic threat. Foxnews.com, May 19. http://www.foxnews.com/story/0,2933,157016,00.html

20. Hoffman, A. H. 1971. *Steal This Book*. New York: Pirate Editions.

21. Ginsberg, Mark, Vince Hoch, and Dan Kroll. 2005. Terrorism and the security of water distribution systems. *Swords and Ploughshares*. January.

22. United States Geological Survey. Water science for schools: Public water supply use. http://ga.water.usgs.gov/edu/wups.html

23. *The 9/11 Commission Report*. 2004. New York: W. W. Norton and Company.

24. Hoover, J. E. 1941. Water supply facilities and national defense. *Journal of the American Water Works Association*. 33 (11):1861.

25. Hickman, Donald C. 1999. A chemical and biological warfare threat: USAF water systems at risk. Counter Proliferation Paper No. 3. USAF Counter Proliferation Center, Air War College.

2

A History of Attacks on Water Supplies

History, despite its wrenching pain, cannot be unlived,
but if faced with courage, need not be lived again.

—Maya Angelou

Those who cannot learn from history are doomed to repeat it.

—George Santayana

Introduction

The compromising of water supplies by adulteration with chemical or biological hazards is nothing new. Intentional attacks on water supplies have a long history, stretching back to ancient times. Some of these attacks were very successful in achieving their goals, while others were thwarted or failed owing to poor planning. The following chronology features those attacks that have been reported. Undoubtedly, there have been many more such events that were unsuccessful, were not recognized at the time as an intentional attack, or were not reported for various reasons.

Chronology of Water Contamination Events

1000 BC. This is the earliest recorded use of intentional water contamination. It has been reported that ancient Chinese warriors used arsenic to contaminate the water supplies of their enemies.[1] In fact, it appears that the ancient Chinese were quite adept at chemical warfare. Ancient writings from this period contain hundreds of recipes for the manufacture of toxic or irritating gases and smokes for use in warfare. One account describes an arsenic-based toxic gas called "soul-hunting fog."[2]

600 BC. *Claviceps purpurea* is a fungus that grows on grasses such as rye. It replaces the grain seed heads with a purplish fungal growth. This fungus contains mycotoxins and is commonly known as rye ergot. The toxins contain alkaloids, which are similar in nature to lysergic acid diethylamide (LSD).[3] The ancient Assyrians placed these infected seed heads in their enemies' wells. Ergot can be very toxic and can produce hallucinogenic symptoms similar to LSD. Large doses of ergot poison can cause delusions, paranoia, uncontrollable twitching, seizures, and cardiovascular problems leading to gangrene and potentially death.[4]

Fig. 2–1. Solon of Athens used the poisoning of water to great advantage during the siege of Cirra.

590 BC. Solon of Athens was said to have used poison during the siege of Cirra, the port city of Delphi in ancient Greece (fig. 2–1). A dispute arose as to the ownership of some land that had been set aside for the god Apollo's temple. Solon was called on to settle the matter and promptly laid siege to the city. Sieges being generally long and drawn-out affairs, Solon decided to speed matters up by placing a dam across the Plesitus River, which was the city's main water supply. The inhabitants, not being ready to submit, relied on a few wells and rainwater to tide them over. Even with these alternate sources, water was in short supply in the city. This is when Solon had his big idea. He is purported to have used a purgative extracted from the roots of the hellebore (commonly known as skunk cabbage) to poison the water behind his makeshift dam. He then broke the dam and allowed the contaminated water to flow into the city. The inhabitants, thinking the siege was coming to an end, gratefully drank their fill of the fresh flowing water. The result of drinking the contaminated water was extreme diarrhea, rendering the defenders of the city unable to resist (sitting down on the job, as it were) when Solon mounted a fresh attack, ending with the city's fall.[3,4]

300 BC. Numerous examples exist of the use of dead animals for the contamination of wells and other water sources in Persian, Greek, and Roman records from this period.[5]

128 BC. The Roman jurists of this time had a saying, *Armis bella non venenis geri*, which translates as "War is fought with weapons, not with poisons." This prescription was not enough to prevent their generals from poisoning wells when it was expedient. Aristonicus was a rebel who had a dispute with the Romans over the control of territory in Asia Minor. His forces were quickly defeated; however, they were not completely destroyed, and Rome was faced with the prospect of a long guerrilla war. To prevent this, they poisoned the local wells that the rebels needed for water supplies. This resulted in a quick Roman victory and the return of Roman rule to the area.[6]

Fig. 2–2. During the battle of Tortona, Barbarossa used human corpses to contaminate wells.

AD 1155. At this time, the Holy Roman Emperor Fredrick I, also known as Barbarossa, invaded northern Italy. During the battle of Tortona, to hasten the fall of the city and to expedite his advance into Italy, he decided to resort to a "new" tactic. He had human corpses placed in the enemy's water supply (fig. 2–2). According to all reports, this was a success, and the water supply was contaminated.[7]

1466. Vlad Tepes, also known as Vlad the Impaler or Vlad Dracula (the basis of the Dracula vampire myth), was engaged in a war with the Ottoman Turks. The location of Transylvania (between Europe and the Ottoman invaders) made him a key player in the defense of Europe from the invading Turkmen. His armies killed so many Turks that the Sultan Mehmed II personally laid siege to Targoviste, one of Vlad's strongholds. Vlad escaped the siege, but left impaled corpses of Muslims to welcome the Turks and used a scorched-earth policy to prevent pursuit. He ordered the burning of crops, the poisoning of wells, and the killing of all animals, so that the Turks would find nothing to eat or drink (fig. 2–3). He set free criminals and encouraged those who were afflicted with leprosy and the plague to mingle among the Turks. Mahmud Pasha lamented, "For six leagues not a drop of water was to be found. The intensity of the heat caused by the scorching sun was so great that the armor seemed as if it would melt like a lighted candle."[8]

Fig. 2–3. Vlad Tepes (Dracula) poisoned wells to prevent pursuit by the Ottoman Turks.

1759. During the Machandal slave rebellion in Haiti, slaves with knowledge of herbs and their properties were able to poison the water supplies of a few planters and animals.[9]

1863. While retreating through Mississippi, Confederate general Joe Johnson ordered the contamination of water sources by placing the carcasses of dead sheep and pigs in wells and ponds. This was to slow the Union advance by depriving them of water sources to allow the Confederates time to regroup. By contrast, the Union Army's General Order 100 stated, "The use of poison in any manner, be it to poison wells, or food, or arms, is wholly excluded from modern warfare."[7]

1904. The United Nations' Whitaker Report of 1985 recognized the German attempt to exterminate the Herero and Nama peoples of Southwest Africa in 1904 as one of the earliest attempts at genocide in the 20th century. In total, approximately 65,000 Herero (80% of the total Herero population) and 10,000 Nama (50% of the total Nama population) were killed or perished. Characteristics of this genocide included death by starvation and the poisoning of wells used by the Herero and Nama populations trapped in the Namib desert. The German general responsible was Lothar von Trotha.[10]

1939. The Japanese military poisoned Soviet water sources with intestinal typhoid bacteria in an area near the former Mongolian border.[11]

1945. Desperate, the retreating German army poisoned a large reservoir in Bohemia with raw sewage.[11]

1946. In the final days of World War II, a group of Jewish resistance fighters hatched a plot to repay the Nazis for the deaths of Jews in the Holocaust. The plan was to poison the water supplies of several major cities in Germany. The goal was to inflict over six million casualties. After much planning, the design was altered because they were unable to stockpile adequate poisons. They eventually infiltrated a bakery that serviced Stalag 13, a German prisoner-of-war camp near Nuremburg. They successfully poisoned three thousand loaves of bread, resulting in the illness of over two thousand German prisoners.[12]

1965. Yasir Arafat's Fatah Movement failed in its first bombing attempt, the bombing of the Israeli National Water Carrier.[13]

The Fatah Movement was responsible for several attacks inside Israel, including attacks on water pipes.[13]

1968. During the Democratic National Convention, in Chicago, members of the Youth International Organization (Yippies), an anti-Vietnam war group led by Abbe Hoffman and Jerry Rubin, threatened to contaminate the city's water supplies with LSD. While most experts didn't take the threat seriously, it contributed to Mayor Richard J. Daley's calling in 7,500 U.S. Army troops and 6,000 National

Guard to back up his 12,000 police officers during the convention. He made sure that for the duration of the convention the city's reservoir was surrounded by armed guards.[14–16]

1970. It was alleged that the water supply of a thousand-acre farm owned and operated by a group of Black Muslims was poisoned, resulting in the death of 30 cows. According to the manager of the farm, the poison appeared as a pinkish-white material found on and around rocks in the stream, which was identified by a local veterinarian as cyanide. Reports indicate that the local KKK might have been responsible.[17,18]

The Weathermen, a radical group active from 1969 to 1976, were a self-described "revolutionary organization of communist men and women" that advocated the overthrow of the government of the United States and the system of capitalism by violent means. In 1970, an informant revealed to U.S. Customs that the Weathermen had plans to steal a biological weapon from Fort Detrick, in Maryland. The plan was to blackmail a homosexual officer at the U.S. Army's bacteriological warfare facility in Fort Detrick into supplying organisms that would then be used to contaminate the water supply of a city or cities in the United States. According to one source, the terrorists apparently succeeded in gaining the cooperation of the officer in question, but "This plot was discovered when the officer requested issue of several items unrelated to his work."[19–22]

1972. In Chicago, several members of a neo-Nazi terrorist organization known as the Order of The Rising Sun (RISE), whose goal was the creation of a new master race, were arrested while holding detailed plans to contaminate the water supplies of several Midwestern cities, including Chicago and Saint Louis. They were in possession of several biological agents that had been produced in the laboratory of a local college. They had botulism, meningitis, anthrax, and up to 40 kg of typhoid cultures.[19–22]

A threat to contaminate New York City's Kensico Reservoir with nerve gas was taken seriously by the FBI. Panicked city officials were nearly ready to declare a health emergency before being calmed by the Army, who assured them that the vast amount needed to contaminate the entire reservoir was impractical.[22,23]

1973. In Germany, a trained biologist threatened to use anthrax and botulinum to contaminate water supplies. The purpose of the threat was blackmail, and the biologist demanded a ransom of $8.5 million to call off the attack.[22,24,25]

1977. In North Carolina, a reservoir was intentionally contaminated. Safety caps were removed, and the water was successfully contaminated, rendering it unusable. Water had to be trucked in for use by local residents.[22,23]

1979. Two separate incidents of chemical poisoning caused 2,008 illnesses in Virginia and Oregon.[20]

1980. An attempt was made to extort money from a Lake Tahoe casino by threatening to contaminate the casino's water supply.[20,21,26]

A disgruntled employee deliberately contaminated water mains in Pittsburgh by injecting weed killer into fire hydrants.[20,21,26]

1982. The Los Angeles Police Department and the FBI arrested a man preparing to poison the water supply of Los Angeles with a biological agent.[27]

1983. After a threat to poison the water supply in Louisiana, traces of cyanide were found.[20,21,26]

In May, the Israeli Government reported that it had uncovered a plot by Israeli Arabs to poison the water supplies in Galilee with an unidentified white powder.[22]

1985. On April 1, an anonymous letter was sent to officials in New York City. The letter threatened to contaminate the city's water supplies with substantial quantities of plutonium unless all charges were dropped against Bernard Goetz, who was on trial for the vigilante shooting of four teenagers whom he alleged were threatening him on a Manhattan subway train in December 1984. Subsequent testing for plutonium revealed that levels were at 200 times the normal concentration for the city's water, but this was still only 0.4% of the level deemed dangerous by the EPA and thus was not a threat to human health.[28]

1986. The Covenant, Sword and Arm of the Lord (CSA) was a Christian Identity survivalist group founded by James Ellison. In 1978, Ellison had a vision of the race war that he believed would soon engulf America, and he transformed his retreat into a white supremacist paramilitary training camp dedicated to the principles of Christian Identity. According to Ellison, the CSA would be an "Ark for God's people" during the coming race war. By "God's people," Ellison meant white Christians. Jews, he told his followers, were not really God's chosen people, but rather a demonic and inferior race.[29] In a raid on the heavily armed camp, the FBI recovered large amount of potassium cyanide. Ellison had plans to use the chemical to poison the water supplies of several major U.S. cities.[30]

1987. In the Philippines, 19 police recruits died, and about 140 were hospitalized after accepting water and sweets from unknown persons.[22]

1990s. A series of reports from unnamed sources indicate a looming, credible threat of attack using VX, an extremely toxic militarized nerve agent similar to sarin, on the water supplies of the nation's capital. These threats were taken so seriously that the government commissioned and funded research into a dedicated online VX detector. Reports indicated that several detectors were built and installed in various secret key locations.

1991. In January, an anonymous letter was sent to the officials of Kelowna, British Columbia. The letter threatened the city's water supplies with biological contaminants. The motive was associated with the Gulf War. Security for the city's water supplies were increased, but no perpetrators were ever identified.[22]

1992. On March 28, there was an attack against the Turkish air force base in Istanbul. The water tanks on the base were found to contain large concentrations of potassium cyanide (50 mg/L). The contamination was discovered before anyone was injured. The Kurdistan Workers Party (PPK) was responsible.[31]

1993. In February, a meeting of Islamic Fundamentalist groups was held in Tehran. At this meeting, under the guidance of the Iranian Foreign Ministry, one of the proposals discussed was the idea of poisoning the water supplies of major cities in the West as a response to aggression against Islamic organizations and states.[32]

1994. In July, a Moldovan general by the name of Nikolay Matveyev threatened to contaminate the water supply of the Russian 14th Army Group in Tiraspol, Moldova, with mercury. He was said to be in possession of 32 kg of mercury. After his dismissal, the mercury was not recovered.[22]

1999. In June, three threatening letters were received by public figures in the United Kingdom. The letters demanded the withdrawal of British troops from Northern Ireland by June 16. The author, Adam Busby, threatened to poison the United Kingdom's water supply with the herbicide paraquat if the British government did not comply with his demand.[33]

Two juveniles poured a bright red substance into the water supply of the town of Grass Valley, California. The water treatment plant was forced to shut down, causing a denial of service to the area's 2,300 residents.[34]

1999–2000. Vitek Boden, a man in his late 40s, applied for a job with the Maroochy Shire Council (in Queensland, Australia), after apparently walking away from a "strained relationship" with Hunter Watertech, an Australian firm that installs supervisory control and data acquisition (SCADA) radio-controlled sewage equipment. The council decided for unknown reasons not to hire him. Boden decided to get even with both the council and his former employer. He packed his car with stolen radio equipment attached to a (possibly stolen) computer. He drove around the area on at least 46 occasions from February 28, 1999, to April 23, 2000, issuing radio commands to the sewage equipment he (probably) helped install. It appears Boden's 46 systematic attacks managed to spill millions of liters of raw sewage. He managed to pollute local parks, rivers, and even the grounds of a Hyatt Regency hotel. One attack accounted for most of his success, requiring a week of cleanup at a cost of AUS$13,110.77 (roughly US$26,000). Boden was caught only when a police officer pulled him over for a traffic violation after one of his attacks. A judge sentenced him to two years behind bars and ordered him to reimburse the aforementioned major cleanup.[35]

2000. On January 23, it was reported that Chechen rebels planned to poison unknown water sources in Chechnya, Russia, to harm Russian Federal forces.[36]

In May, the Anatolia news agency reported that a man was arrested for attempting to poison the water supply of the village of Kurusaray, Turkey, with insecticide.[36]

In June, Palestinian news reported that Israeli settlers from the Efrat settlement had deliberately released sewer water into Palestinian agricultural fields in the village of Khadder, in the West Bank.[36]

On September 15, a day after residents in a condominium block in Singapore had complained of a strange odor in their water supply, it was discovered that the water tank had been deliberately poisoned with kerosene and turpentine.[36]

2001. In Ladysmith, British Columbia, workers found a reservoir hatch open and an oily substance on the surface of the water. Hatches had been removed at two other nearby facilities earlier in the year, but no contamination was noted. The 6,400 residents of Ladysmith were warned not to drink the water until the system could be flushed.[37]

In the Philippines, Abu Sayyaf, an Islamic extremist organization associated with al Qaeda, threatened to poison the water supply of the Christian town of Isabella. About a week later, the water supply to several villages in the same area of the Philippines was cut off when residents complained of a gasoline taste and odor in the water. Abu Sayyaf was blamed.[38]

2002. In February, al Qaeda members were arrested in Rome, in the process of orchestrating an attack on the water distribution system around the U.S. Embassy. They were widely reported to be in the process of contaminating the system with cyanide when they were arrested. They had detailed plans and equipment to carry out the attack and were thwarted at the last moment. The compound they were in possession of turned out, however, to be a relatively benign cyanide derivative, potassium ferricyanide.[39] See chapter 3 for a more detailed discussion of this event.

In February, FARC rebels poisoned a water treatment plant in the town of Pitalito, Colombia. The substance, which was not identified, was detected during a routine water test.[40]

In June, workers at a water utility in Janesville, Wisconsin, found that the barbed wire of a perimeter fence and the padlock on a five-million-gallon storage tank had been cut. There was no direct evidence of contamination, but the tank was drained and superchlorinated as a precaution.[41]

In July, federal officials in Denver arrested two al Qaeda terror suspects in possession of documents about how to poison the country's water supplies.[42]

Also in July, Russian special services participating in counterterrorist operations in the northern Caucasus uncovered information that Chechen rebels were planning to use a potent substance to poison water and food supplies in Grozny.[40]

In November, Boere Vryheids Aksie (BVA), a South African White Supremacist group, planned to poison millions of black South Africans by contaminating water supplies with tetranium, an agricultural poison.[43]

In December, al Qaeda operatives were arrested with plans to attack water supply networks surrounding the Eiffel Tower neighborhoods of Paris.[44] This was widely unreported in the French press owing to the French government's lack of support for the United States preceding the Iraq war.

2003. In January, at a water treatment plant located in Debary, Florida, someone crossed a barbed wire fence, broke the lock on an entry gate, and removed the screens from the aerator. This had the indications of an insider or professional attack. The damage could have affected the water supply for four thousand homes. The utility was assessed a fine for failing to promptly notify the Health Department of the break-in, as required by statute.[45]

In April, Jordan foiled an Iraqi plot to poison drinking water from the city of Zarqa, supplying U.S. military bases along the eastern desert.[46]

In September, an FBI bulletin warned of al Qaeda plans found in Afghanistan to poison U.S. food and water supplies.[47]

In October, a vial containing the deadly poison ricin was found in a Greenville, South Carolina, postal facility. Accompanying the vial was a note stating that the city's water supply would be contaminated with ricin unless certain demands were met. These demands were proposed changes to federal policy as it pertained to the number of hours that overland truckers were allowed to drive without resting. Subsequent testing revealed that there was no ricin in the water.[48]

In China, a man was arrested after dumping a pint of insecticide into a reservoir. After exposure, 64 people became ill, and 42 were hospitalized. The perpetrator was a seller of water purification devices, and his only motive was to increase sales.[49]

2004. In April, a Sudanese man with Iranian intelligence contacts was captured carrying a very powerful poison in Iraq. The man was reportedly preparing to poison the water supply of Diwaniyah, a city 100 miles south of Baghdad.[50]

The FBI and the Department of Homeland Security (DHS) issued a bulletin warning of terrorists trying to recruit workers in water plants, as part of a plan to poison drinking water at the treatment plants.[51]

2005. The Sudanese government and their Arab allies poisoned wells in attacks on black civilians in the Darfur region. Representative Henry Hyde, Chair of the Committee on International Relations, questioned the U.N. Commission of Inquiry's findings that the actions in Darfur did not constitute genocide. He particularly sited the poisoning of wells as evidence of intent to commit genocide.[52]

According to the *Los Angeles Times,* before the celebration, attended by President Bush, of the 60th anniversary of the end of World War II in Europe, Russian authorities recovered a cache of poisons intended for use by terrorists to disrupt the celebrations. Liquid cyanide and other unidentified poisons were discovered.

The Russian Federal Security Services said, "The use of these strong-acting poisons in small doses in highly populated areas, in key installations and in reservoirs could have caused large numbers of victims."[53]

Rumors attempting to disrupt the second round of Iraqi parliamentary elections swept Baghdad on December 15 that the water supply had been poisoned only hours before polls were to open. Residents were awakened at 1 a.m. when warnings about drinking water were broadcast through mosque loudspeakers. According to the Associated Press, the country's health minister, Abdel Mutalib Mohammed, issued a statement on television saying that there were no cases of poison in the water system and that the news was untrue.[54]

Conclusion

As these examples clearly demonstrate, the intentional poisoning of water supplies is nothing new and has by no means been rare throughout history. That water supplies are a target is also reinforced by continued interest of domestic terrorist and fringe groups in using a chemical, biological, or radiological (CBR) agent in their attacks. Islamic extremist groups have also exhibited interest in water supply systems, as substantiated by the more recent events detailed in the chronology.

Some experts have maintained that Islamic extremists would be highly unlikely to attack water, because it is held to be sacred by their religion. Like all beliefs, the ones followed by Islamic extremists are open to individual interpretation, and justification for attacking water can be found, if that is what someone is looking for. Several passages from Islamic writings that could be construed to justify attacks on water follow.

> Hell is before him, and he is made to drink a festering water, which he sippeth but can hardly swallow, and death commeth unto him from every side while yet he cannot die and before him is a harsh doom.
>
> —Abraham 14:16–17

> Those who disbelieve will be forced to drink boiling water, and will face a painful doom.
>
> —The Cattle 6:70

> Those who deny the Scripture and Allah's messengers will be dragged through boiling water and thrust into the Fire.
>
> —The Believer 40:70–72

> Those in the Garden will drink delicious wine, while those in the Fire will drink boiling water that will tear apart their intestines.
>
> —Muhammad 47:15

These examples are not meant to infer that radical Islamic organizations are any more likely to attack water than any other terrorist organization. They are just meant to show that if they had this mode of attack in mind, there are ways that it could be justified. We see examples of this with other terrorist organizations, such as environmental groups that cause environmental damage in their attacks and animal rights groups that cause animals to suffer by setting them free in environments in which they are not prepared to survive. If the goal is to orchestrate an attack, then justification for the mode of attack can always be rationalized.

While recent attempts on water have been thwarted or unsuccessful, history shows that al Qaeda and other diligent extremist groups have a unique ability to refine and perfect attack strategies. For example, when al Qaeda made their first attempt to topple the World Trade Center, they were unsuccessful. After several years of planning and adjusting their mode of attack, we know that they were successful on their second try. Just because they have not been successful in their water operations doesn't mean that they will abandon their plans in this area. They are more likely to make operational changes that they feel will lead to success in these endeavors.

The threat posed to drinking water is as important to key infrastructure targets—in particular, U.S. military bases—as it is to private and government "icon" facilities and general water supply networks. The policy requiring that U.S. military bases use off-base water sources whenever possible adds to their vulnerability. If in a time of war it were an enemy's goal to disable a specific base—for instance, an air base used to conduct military operations against an overseas target—it could be attacked in such a way as to make the personnel unfit to perform their mission. Something as simple as a bad case of diarrhea may be enough to fulfill this goal.

Another example would be military supply bases. If the personnel at a supply base were disabled so that they couldn't load or unload cargo, it would have a rapid adverse effect on field operations that rely on those supplies. Researchers from the U.S. Air Force, the Army Corps of Engineers, and Hach Homeland Security Technologies

(HST) have calculated that an attack on drinking water distribution systems could be mounted for between $0.05 and $5.00 per death, using rudimentary techniques, and could amass casualties in the thousands over a period of hours.[55–58]

Recent events, such as the Rome cyanide incident, may give us confidence that the terrorists, in fact, don't know what they are doing. That they used a compound (potassium ferricyanide) that cannot be called toxic leads us to believe that these groups are not technologically up to the task of contaminating a water supply. A closer examination of the Rome incident in the next chapter reveals a more sinister interpretation of the facts involved in the case and should help to dispel this idea of incompetence on the part of al Qaeda and other terrorist organizations or individuals.

Notes

1. Di Rado, Alicia. Understanding the tools of terror. University of Southern California Public Relations News Room. http://www/usc.edu/uscnews/stories/8890.html
2. Wikipedia. Chemical warfare. http://en.wikipedia.org/wiki/Chemical_warfare
3. Tschanz, David W. "Eye of newt and toe of frog": Biotoxins in warfare. Strategypage.com. http://www.strategypage.com/articles/?target=biotoxin.htm&reader=long
4. Texas Department of State Health Services, Pubic Health Preparedness. Bioterrorism: History of bioterrorism. http://www.tdh.state.tx.us/preparedness/bioterrorism/public/history/
5. E Medicine and Consumer Health. Biological warfare. http://www.emedicinehealth.com/articles/15704-1.asp
6. A Brief History of Chemical and Biological Weapons. http://www.cbwinfo.com/History/History.html
7. Iserson, Kenneth. 2002. Viruses and vivisections—Japan's inhuman experiments: The history of biological warfare. In *Demon Doctors: Physicians as Serial Killers.* Tucson, AZ: Galen Press.
8. Wikipedia. Vlad II Dracula. http://en.wikipedia.org/wiki/Vlad_III_Dracula
9. The Haitian Independence Movement. http://phuhs.org/academics/IB/Burton/examreview_HTML/independence_movements2002.htm
10. Nation Master Encyclopedia. Genocides in history. http://www.nationmaster.com/encyclopedia/Genocides-in-history
11. Arizona Department of Health Services, Division of Public Health Services. History of biowarfare and bioterrorism. http://www.azdhs.gov/phs/edc/edrp/es/bthistor2.htm
12. Davis, Douglas. 1998. Survivor reveals 1945 plan to kill 6 million Germans. *The Jewish News Weekly of Northern California*. March 27.

13. Committee for Accuracy in Middle East Reporting. Yasir Arafat's timeline of terror. http://www.camera.org/index.asp?x_context=7&x_issue=11&x_article=795
14. Giordano, Al. Chicago 96. http://www.bostonphoenix.com/alt1/archive/news/CHICAGO_96.html
15. The History Channel. Jerry Rubin: Counterculture figure addresses the Yippie Convention. http://www.historychannel.com/speeches/archive/speech_265.html
16. Linder, D. O. The Chicago Seven conspiracy trial. http://www.law.umkc.edu/faculty/projects/ftrials/Chicago7/Account.html
17. Wooton, James. 1970. Poison is suspected in death of 30 cows on a Muslim farm. *The New York Times.* March 16.
18. Wooton, James. 1970. Black Muslims would sell farm to Klan. *The New York Times.* March 17.
19. Mengel, R. W. 1976. Terrorism and new technologies of destruction: An overview of the potential risk. In *Disorders and Terrorism,* 443–473. Washington, DC: U.S. Government Printing Office.
20. Terrorism Research Group. n.d. *Chronology of Chemical-Biological Incidents.* Santa Monica, CA: Rand.
21. Falkenrath, R. A., R. D. Newman, and B. A. Thayer. 1998 *America's Achilles' Heel: Nuclear, Biological and Chemical Terrorism and Covert Attack.* Cambridge, MA: MIT Press.
22. Purver, R. 1995. *Chemical and Biological Terrorism: The Threat According to the Open Literature.* Ottawa: Canadian Security Intelligence Service.
23. Clark, Richard Charles. 1980. *Technological Terrorism.* Old Greenwich, CT: Devin-Adair.
24. Jenkins, Brian M., and Alfred P. Rubin. 1978. New vulnerabilities and the acquisition of new weapons by nongovernment groups. In *Legal Aspects of International Terrorism.* Lexington, MA: Lexington Books.
25. Kupperman, Robert H., and Darrell M. Trent. 1979. *Terrorism: Threat, Reality, Response.* Stanford, CA: Hoover Institution Press.
26. Ownbey, P. J., F. D. Schaumburg, and P. C. Klingeman. 1988. Ensuring the security of public water supplies. *Journal of the American Water Works Association.* 80 (2): 30–34.
27. Payne, Carrol. 2003. Understanding terrorism—weapons of mass destruction. *World Conflict Quarterly.* September. http://www.globalterrorism101.com/UTWMD.html
28. *Time Magazine.* 1985. American notes. August 5.
29. MIPT Terrorism Knowledge Base. http://www.tkb.org/Group.jsp?groupID=3226
30. Mullins, Wayman C. 1992. An overview and analysis of nuclear, biological, and chemical terrorism: The weapons, strategies and solutions to a growing problem. *American Journal of Criminal Justice.* 16 (2): 95–119.
31. Chelyshev, Alexander. 1992. Terrorists poison water in Turkish Army cantonment. *Telegraph Agency of the Soviet Union (TASS).* March 29.

32. Haeri, Safa. 1993. Iran: Vehement reaction. *Middle East International.* March 19.

33. Camern, Gavin, Jason Pate, Diana McCauley, and Lindsay DeFazio. 1999. WMD terrorism chronology: Incidents involving sub-national actors and chemical, biological, radiological, or nuclear materials. Center for Nonproliferation Reports, Monterey Institute for International Studies. http://cns.miis.edu/pubs/npr/vol07/72/wmdchr72.htm

34. Cox, J. 1999. Vandals pollute Grass Valley water supply: Two teens held. *Sacramento Bee*. October 13.

35. Smith, Tony. 2001. Hacker jailed for revenge sewage attacks. *The Register*. October 31. http://www.theregister.co.uk/2001/10/31/hacker_jailed_for_revenge_sewage/print.html

36. Pate, Jason, Gary Ackerman, and Kimberly McCloud. 2000. WMD terrorism chronology: Incidents involving sub-national actors and chemical, biological, radiological, or nuclear materials. Center for Nonproliferation Reports, Monterey Institute for International Studies. http://cns.miis.edu/pubs/reports/cbrn2k.htm

37. *Canadian Press*. 2001. Vancouver Island town declares emergency after water supply tampered with. http://teamalberta.net/water/010728a.html

38. Reuters. 2001. Philippine rebels suspected of water "poisoning." October 16. http://www.planetark.com/dailynewsstory.cfm/newsid/12807/story.htm

39. BBC News. 2002. Cyanide attack foiled in Italy. February 20. http://news.bbc.co.uk/1/hi/world/europe/1831511.stm

40. Turnbull, Wayne, and Praveen Abhayaratne. 2002. WMD Terrorism Chronology: Incidents Involving Sub-National Actors and Chemical, Biological, Radiological, or Nuclear Materials. Center for Nonproliferation Reports, Monterey Institute for International Studies. http://cns.miis.edu/pubs/reports/pdfs/cbrn2k2.pdf

41. Carpenter, C. 2002. Security threat a dress rehearsal for Janesville. *Opflow*. 28:7.

42. Cameron, Carl. 2002. Feds arrest al Qaeda suspects with plans to poison water. *Fox News*. July 30. http://www.frwa.net/ARTICLES/feds_arrest_al_qaeda_suspects_wi.htm

43. Basildon, Peta. 2002. Rightwing plot to poison water foiled. *Independent On-Line*. November 25. http://www.iol.co.za/index.php?click_id=6&art_id=ct20021125111353852R235535&set_id=1

44. Von Derschau, V. 2002. Radicals arrested near Paris over poison gas attack plot. Associated Press. December 18.

45. Voyles-Pulver, D. 2003. State calls water break-in "professional." *Daytona Beach News Journal*. January 23.

46. Feuer, A. 2003. Jordan Arrests Iraqis in Plot to Poison Water. *New York Times*. April 2.

47. CBS News. 2003. Al Qaeda might use poison. September 5. http://www.cbsnews.com/stories/2003/09/05/national/main571778.shtml

48. Smith, Tim. 2003. South Carolina ricin not terrorism, officials say. *The Greenville News*. October 21.

49. BBC News. 2003. China salesman poisoned water. October 6. http://news.bbc.co.uk/2/hi/asia-pacific/3167122.stm

50. *Turkish Press*. 2004. Iraqi defense minister warns Iran over sending of spies and saboteurs. July 26. http://www.turkishpress.com/news.asp?id=23168

51. Kalil, J. M., and Dave Berns. 2004. Drinking supply: Terrorist had eyes on water. *Las Vegas Review Journal*. August 12. http://www.reviewjournal.com/lvrj_home/2004/Aug-12-Thu-2004/news/24519286.html

52. U.S. House of Representative, Committee on International Relations. 2005. Hyde questions findings of UN Commission of Inquiry on Darfur. Press release. http://www.house.gov/international_relations/109/news020105.htm

53. Murphy, Kim. 2005. Poison, explosives found before world leaders' visit, Russia says. *Los Angles Times*. May 6.

54. Associated Press. 2005. Poison water rumors during Iraq vote. *Newsday*. December 15.

55. Kroll, Dan. 2003. Mass casualties on a budget. Confidential paper. Hach HST.

56. Army Corps of Engineers. n.d. Calculations on threat agents and requirements and logistics for mounting a successful backflow attack.

57. Allman, T. P. 2003. Drinking water distribution system modeling for predicting the impact and detection of intentional contamination. Masters thesis. Department of Civil Engineering, Colorado State University.

58. Ginsberg, Mark, Vince Hoch, and Dan Kroll. 2005. Terrorism and the security of water distribution systems. *Swords and Ploughshares*. January.

3

A Reevaluation of the Rome Incident: Don't Underestimate the Enemy

Never underestimate the other guy.

—Jack Welch

On February 20, 2002, at 3:39 p.m., the Associated Press released a story with the following headline, "4 arrested with cyanide and Rome water supply maps."[1] The article detailed a raid on an apartment and the arrest of four Moroccan terrorists. They were part of the Salafist Group for Call and Combat, an Algerian organization with ties to Osama bin Laden and the al Qaeda network. According to the article, they were in possession of a common cyanide-based industrial chemical later identified as potassium ferricyanide, as well as false documents and detailed maps of the area surrounding the U.S. Embassy. The maps found in the apartment included details of the city's water system, and the U.S. Embassy was circled. Investigators believed the suspects planned to contaminate the water supplies in the capital, including the commercial area around Via Veneto, where the U.S. Embassy is located (fig. 3–1).[2]

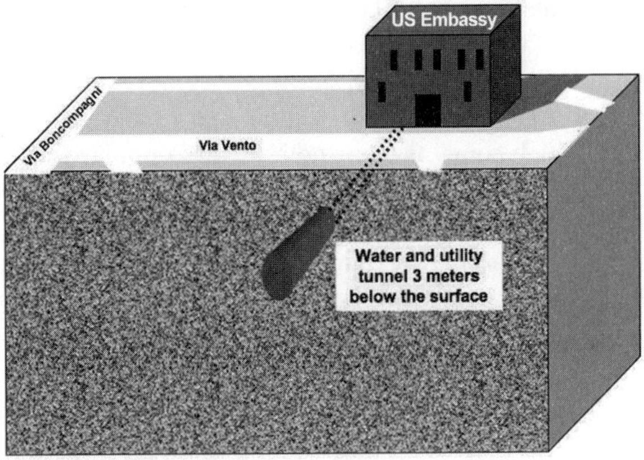

Fig. 3–1. The U.S. Embassy and the utility tunnel used to gain access to the water pipes

After this story became public, most people in the water and security businesses breathed a sigh of relief. The typical reaction was, "Thank goodness we were dealing with such inept terrorists. They were trying to poison a water system with a compound that isn't even really poisonous." The LD_{50} (the amount needed to kill 50% of those exposed) for potassium ferricyanide is 2,970 mg/kg of body weight, while that of sodium cyanide is 6.4 mg/kg (fig. 3–2). Therefore, if you were attempting to poison someone with ferricyanide, you would need 464 times the amount of sodium cyanide that would be required. It would take over 200 grams of ferricyanide to kill the average adult male. This represents almost a full cup of the compound, which is bright orange. Even the most unobservant victim would notice someone trying to slip this into their water. In fact, the toxicity of potassium ferricyanide is so low that it isn't much more toxic than common table salt, sodium chloride. The LD_{50} for sodium chloride is 3,750 mg/kg of body weight (fig. 3–2).[3]

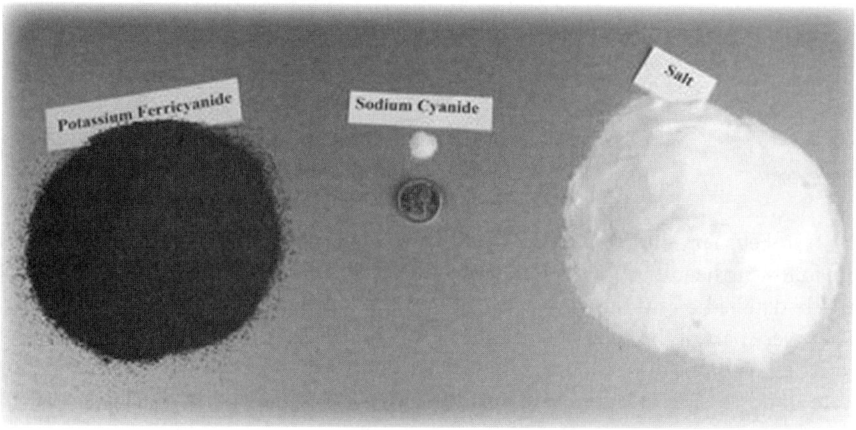

Fig. 3–2. Volumes of material that it would take to kill 50% of adult males exposed (with a U.S. quarter for size reference). Table salt would be as effective a poison as potassium ferricyanide.

Even though it is tempting to say, "Wow! It is a darned good thing that these guys don't know what they are doing," on further reflection, this interpretation of the events doesn't make any sense. According to all reports, these terrorists did know what they were doing. They had chosen the perfect location to orchestrate an attack on the embassy. The pipes were chosen and accessed correctly. There was also a large municipal water tank nearby that served the surrounding Via Veneto area, which contains many hotels and is popular with upscale tourists. A successful attack would have reached the embassy compound and resulted in many casualties in the surrounding area. So why did they make such a huge mistake as to choose the wrong chemical? Five minutes on the Internet, and a high school education would have pointed out the mistake. A search for "ferricyanide" on Google yields over 80,000 hits.

Documents dealing with chemical attacks have been captured in Afghanistan, in Iraq, and within the United States and demonstrate that the terrorists do not lack knowledge of toxicology. They seem to have a sophisticated knowledge of what is required in order to cause harm. It is extremely unlikely that they would make such a simple mistake as using ferricyanide rather than a toxic cyanide compound, especially when toxicological data for these and a huge variety of other readily available compounds are so easily obtained from the public literature.

Instead, it is more likely that the use of ferricyanide was intentional. Why would they use a nontoxic chemical? There is one answer to this question that makes sense. While it is easy to determine what chemicals you could use to poison a water system through a simple Internet search and a rudimentary understanding of chemistry, targeting a specific site, such as the embassy, by introduction of a toxicant into the water distribution system through a backflow event is more difficult. The labyrinth of pipes that compose a water distribution system is very complex, and flow patterns are often difficult to map. Many system operators that have years of experience with a given system do not always have the detailed knowledge it would take to direct a contaminant to a specific site. A good guess as to the motives of the would-be terrorists in the case of the ferricyanide is not a dismal failure in understanding toxicology, but rather a dry run of the attack scenario to determine whether they had correctly mapped the hydraulics of the system.

To garner more knowledge about the flows in their systems, operators often perform tracer studies. This is when they inject a benign compound into a system and track its movement over time. To be a good candidate as a tracer, a compound should have certain characteristics: it should have solubility properties and flow characteristics close to those of the actual compound of interest; it should be relatively nontoxic and easy to work with; it should be easily detectable through simple analytical measurement; and it should be easy to obtain and inexpensive.

While ferricyanide is not toxic, it does have many characteristics in common with such very toxic cyanide compounds as sodium cyanide and potassium cyanide. The solubility and flow characteristics of ferricyanide in water closely mimic those of the other cyanide compounds. It is inexpensive and easy to obtain. Also, solutions of ferricyanide are bright yellow in color, making concentrations of as little as 1 mg/L detectable with the naked eye.[4] Best of all, the color is so light at this level that, unless you are looking for it, it would most likely not be noticeable or would be interpreted as a slight amount of rust in the water. The terrorists could inject a small amount of the ferricyanide into the system and then send a cohort into the embassy on legitimate business who, while there, would observe the drinking or restroom water for traces of the telltale yellow color. All it would take would be a legitimate trip to use the restroom and casual observation as to the color of the water to see whether the attack scenario would work, hitting the intended target. Once the hydraulics were verified, the real attack using a different cyanide compound could commence.

When looking at the newspaper articles in this light, the results are not so reassuring. Public sources of information do not always present the full picture of terrorist capabilities. Making assumptions about the operational effectiveness of terrorist organizations on the basis of the publicly available literature is a mistake that others have made in the past.

According to David Kay, in a chapter entitled the "WMD Threat: Hype or Reality?" in the book "*The Terrorism Threat and U.S. Government Response: Operational and Organizational Factors*," academic analysts made this mistake in regard to the Palestine Liberation Organization (PLO) and the IRA. After an in-depth study of the statistics, the analysts concluded that the PLO and the IRA could not be as technically skilled as the U.S. government asserted, because they were frequently blowing themselves up as they produced and deployed their bombs. What the analysts didn't know was that the government at that time was deploying preventative measures that made sure that a stream of defective bomb-making material was funneled into the terrorists' supply chain and that these prematurely detonated the devices while they were being placed around vulnerable areas. The statistics in the public records were wrong; thus, the conclusions were wrong.[5]

As private citizens we are not privy to the full spectrum of intelligence information that exists. That is why it is prudent for us to take a cautious stance in our interpretation of the data that we do have available. It is far better to err on the side of caution and interpret news and events in a dark light than it is to have a false sense of security in imagining that our enemies are not as clever or as resourceful as they really are.

The interpretation of the Rome incident that I have detailed here is much more likely than the scenario of stupid terrorists. It never pays to underestimate your enemy, and I fear in this instance many people have done just that. The assault on the Rome embassy was not a bungled attempt by amateurs, but a sophisticated experiment that with luck was interrupted before its deadly conclusion. Lord Acton, who is famous for saying that "Power corrupts, and absolute power corrupts absolutely," also had another saying that, while less famous, is very pertinent to the subject at hand: "Do not overlook the strength of the bad cause, or the weakness of the good."[6]

On April 28, 2004, after a two-year trial, an Italian court acquitted all 12 defendants (9 Moroccans, 1 Pakistani, 1 Tunisian, and 1 Algerian) who had been charged as being complicit in the Rome plot. The accused denied any knowledge of the chemicals or documents, saying many people passed through the apartment in which they were discovered. The defendant's lawyers also emphasized in their defense that the cyanide compound was not a dangerous substance and might, at worst, have been used to forge identification documents. Only one defendant was found guilty of a crime, and that was on the minor charge of receiving license plates from a stolen scooter; he was sentenced to a six-month jail term and was fined 100 euros (about US$119).[7] Perhaps if the court had been made aware of the potential that the ferricyanide was used as a tracer, the verdict would have been different.

It is my hope that by this point I have been able to convince you that water is indeed a target of terrorist activity. Such attacks have occurred in the past and continue to occur today. That these attacks have met with little or limited success should not decrease our sense of urgency in addressing this vulnerability. The enemy will learn by their mistakes, and so should we. We face a skilled and determined foe that has as their goal the infliction of damage and massive casualties by any means at their disposal; it behooves us to counter their intentions with any means at our disposal. The remainder of this book will deal with the various modes of attack that terrorists could employ in an assault on our water supplies and with actions that can be taken to counter these potential assaults.

Notes

1. Associated Press. 2002. 4 arrested with cyanide and Rome water supply maps. *Houston Chronicle.* February 20. http://www.chron.com/disp/story.mpl/world/1262805.html

2. BBC News. 2002. Cyanide attack foiled in Italy. February 20. http://news.bbc.co.uk/1/hi/world/europe/1831511.stm

3. Budavari, Susan, ed. 1989. *The Merck Index.* 11th edition. Rahway, NJ: Merck and Company.

4. Results of experiments conducted at Hach HST.

5. Kay, David. 2001. WMD terrorism: Hype or reality? In *The Terrorism Threat and U.S. Government Response: Operational and Organizational Factors.* Edited by James M. Smith and William C. Thomas. INSS Book Series. U.S. Air Force Academy, CO: USAF Institute for National Security Studies.

6. Lord Acton. 1895. Inaugural lecture on the study of history. June 11.

7. Balmer, Crispin. 2004. Italian court acquits nine in alleged plot against U.S. Embassy. *Reuters.* April 29.

4

Water Supply Vulnerabilities: How and Where Could an Attack Occur?

The innocence that feels no risk and is taught no caution, is more vulnerable than guilt, and oftener assailed.

—Nathaniel P. Willis

State of the System

The water supply network in the United States and other countries consists of centuries-old infrastructure components coupled with modern high-tech equipment (fig. 4–1). Because of the aging infrastructure that plagues most municipal water supply systems, drinking water and wastewater infrastructure investment costs over the next 20 years may range from $492 billion to $820 billion, according to a Congressional Budget Office (CBO) report.[1] It is not uncommon to have, in the same system, pipes that are made of wood and were installed over 100 years ago feeding into or from ductile iron and polyvinyl chloride (PVC) pipes all installed within the past 20 years.

The whole system is regularly monitored by ultrasonic flow meters and a computerized SCADA system (for gathering and analyzing real-time data). This sprawling system cobbled together from these various components supplies water for drinking and other domestic purposes, as well as industrial uses. Even though this system currently provides some of the cleanest and safest water anywhere in the world, the very nature of the network as it is configured today lends to it the very vulnerability that could be exploited by those who wish to cause harm.

Fig. 4–1. Much of the drinking water infrastructure is aging and in poor or marginal condition, while other parts of the water supply network are state of the art and have been furbished with modern equipment.

Federal Recognition of the Problem

> The water supplied to U.S. communities is potentially vulnerable to terrorist attacks by insertion of biological agents, chemical agents, or toxins. . . . The possibility of attack is of considerable concern, . . . [and] these agents could be a threat if they were inserted at critical points in the system; theoretically, they could cause a large number of casualties.
>
> —The President's Critical Infrastructure Assurance Office

This vulnerability of the water supply systems to terrorist acts was recognized by the government prior to 9/11. President Clinton issued Presidential Decision Directive 63 (PDD 63), on May 22, 1998. It identified eight critical infrastructures to the United States. The water infrastructure, including drinking water and wastewater industries, was one of the original eight. The directive called for "vulnerability assessments . . . for each sector of the economy and each sector of the government that might be a target of infrastructure attack intended to significantly damage the United States," and "within both the government and the private sector to sensitize people to the importance of security and to train them in security standards."

In 1999, the military took an aggressive look at the problem. In that year, Major Donald C. Hickman described how the mandatory use of local utilities whenever possible by domestic military bases left them open to terrorist or military assaults on their water supplies:

> Clearly, USAF operations could be severely degraded by CW/BW attacks on an air base water system or by attacks on a community water supply that the air base is dependent upon. A center of gravity, drinking water is essential to the airmen who operate and support USAF weapon systems. It is collected, treated, stored, and delivered in systems with common critical points. These points, if vulnerable, are potential CW/BW targets. A terrorist bent on killing Americans, or a rogue nation seeking asymmetric advantage in a pre-emptive strike, has available a plethora of CW/BW agents and materials that are effective in drinking water. Rhetorically, why use ballistic missiles when a thermos laced with cholera or botulism toxin, or a couple of bags of "cement" (sodium cyanide), could functionally destroy operations? As these examples demonstrate, chemicals, pathogens and toxins are cheap, ubiquitous, and deadly in water. Very little material is needed to inflict lethal or incapacitating doses in an unsuspecting and unprepared population. Adversaries seeking asymmetric advantage could focus on the water attack scenario, severing the USAF's proverbial "Achilles' Heel," grounding operations before "wheels are in the well."[2]

The civilian vulnerability to terrorist acts took on increased urgency after 9/11. In June 2002, President Bush signed the Public Health Security and Bioterrorism Preparedness and Response Act (referred to as the Bioterrorism Act). Several provisions in this act dealt directly with the issue of water supply vulnerability. These provisions modified the Safe Drinking Water Act (SDWA). The SDWA now requires community waterworks serving populations of more than 3,300 to conduct vulnerability assessments. Small systems (serving populations between 3,300 and 49,999) had until June 30, 2004, and medium systems (serving populations between 50,000 and 99,999) had until December 31, 2003, to complete vulnerability assessments. Large drinking water systems (serving populations over 100,000) had until March 31, 2003, to submit their vulnerability assessments to the United States Environmental Protection Agency (EPA). Within six months of submitting the vulnerability assessment, the waterworks had to revise or complete an emergency response plan.

The SDWA provided funding for basic security enhancements including but not limited to the following:

- Purchase and installation of equipment for detection of intruders
- Purchase and installation of fencing, gating, lighting, or security cameras
- Tamper-proofing of manhole covers, fire hydrants, and valve boxes
- Rekeying of doors and locks
- Improvements to electronic, computer, or other automated systems and remote security systems
- Participation in training programs and the purchase of training manuals and guidance materials relating to security against terrorist attacks
- Improvements in the use, storage, or handling of various chemicals
- Security screening of employees or contractor support services

Funding was also authorized for other areas of interest. A review of the act gives a good outline of the vulnerable areas in the system:

- Methods and means by which pipes and other constructed conveyances utilized in public water systems could be destroyed or otherwise prevented from providing adequate supplies of drinking water meeting applicable public health standards
- Methods and means by which collection, pretreatment, treatment, storage, and distribution facilities used in connection with public water systems and collection and pretreatment storage facilities used in connection with public water systems could be destroyed or otherwise prevented from providing adequate supplies of drinking water that meets applicable public health standards
- Methods and means by which pipes, constructed conveyances, collection, pretreatment, treatment, storage, and distribution systems used in connection with public water systems could be altered or affected to cross-contaminate drinking water supplies

- Methods and means by which pipes, constructed conveyances, collection, pretreatment, treatment, storage and distribution systems used in connection with public water systems could be reasonably protected from terrorist attacks or other acts intended to disrupt the supply or affect the safety of drinking water
- Methods and means by which information systems, including process controls and supervisory control and data acquisition and cyber systems at community water systems could be disrupted by terrorists or other groups

The basic vulnerabilities recognized were therefore physical destruction of pipes or infrastructure, disabling of treatment, contamination of source or finished water, cyber attack, and insider attack.

Also included in SDWA was funding for contaminant prevention, detection, and response. The areas of interest specified in this part of the act give an indication as to how some vulnerabilities can be eliminated or lessened. To prevent, detect, and respond to the intentional introduction of CBR contaminants in community water systems and in source water for community water systems, SDWA authorizes the EPA administrator to enter into contracts or cooperative agreements to provide for a review of current and future methods including the following:

- Methods, means, and equipment, including real-time monitoring systems, designed to monitor and detect various levels of CBR contaminants or indicators of contaminants and reduce the likelihood that such contaminants can be successfully introduced into public water systems and source water intended to be used as drinking water
- Methods and means to provide sufficient notice to operators of public water systems, and individuals served by such systems, of the introduction of CBR contaminants and the possible effect of such introduction on public health and the safety and supply of drinking water
- Methods and means for developing educational and awareness programs for community water systems
- Procedures and equipment necessary to prevent the flow of contaminated drinking water to individuals served by public water systems
- Methods, means, and equipment that could mitigate a deleterious effect on or negate the public health and the safety and supply caused by the introduction of contaminants into water intended to be used as drinking water, including an examination of the effectiveness of various drinking water technologies in removing, inactivating, or neutralizing CBR contaminants
- Biomedical research into short-term and long-term impacts on public health of various CBR contaminants that could be introduced into public water systems through terrorist or other intentional acts

In 2004, President Bush issued Presidential Homeland Security Directive 9, which also had several sections addressing water:

> (8) The Secretaries of the Interior, Agriculture, Health and Human Services, the Administrator of the Environmental Protection Agency, and the heads of other appropriate Federal departments and agencies shall build upon and expand current monitoring and surveillance programs to:
>
>> (a) develop robust, comprehensive, and fully coordinated surveillance and monitoring systems, including international information, for animal disease, plant disease, wildlife disease, food, public health, and water quality that provides early detection and awareness of disease, pest, or poisonous agents;
>>
>> (b) develop systems that, as appropriate, track specific animals and plants, as well as specific commodities and food; and
>>
>> (c) develop nationwide laboratory networks for food, veterinary, plant health, and water quality that integrate existing Federal and State laboratory resources, are interconnected, and utilize standardized diagnostic protocols and procedures.
>
> (9) The Attorney General, the Secretary of Homeland Security, and the Director of Central Intelligence, in coordination with the Secretaries of Agriculture, Health and Human Services, and the Administrator of the Environmental Protection Agency, shall develop and enhance intelligence operations and analysis capabilities focusing on the agriculture, food, and water sectors. These intelligence capabilities will include collection and analysis of information concerning threats, delivery systems, and methods that could be directed against these sectors.
>
> (10) The Secretary of Homeland Security shall coordinate with the Secretaries of Agriculture, Health and Human Services, and the Administrator of the Environmental Protection Agency, and the heads of other appropriate Federal departments and agencies to create a new biological threat awareness capacity that will enhance detection and characterization of an attack. This new capacity will build upon the improved and upgraded surveillance systems described in paragraph 8 and integrate and analyze domestic and international surveillance and monitoring data collected from human health, animal health, plant health, food, and water quality systems. The Secretary of Homeland Security will submit a report to me through the Homeland Security Council within

> 90 days of the date of this directive on specific options for establishing this capability, including recommendations for its organizational location and structure.
>
> (23) The Secretaries of Homeland Security, Agriculture, and Health and Human Services, the Administrator of the Environmental Protection Agency, and the heads of other appropriate Federal departments and agencies, in consultation with the Director of the Office of Science and Technology Policy, will accelerate and expand development of current and new countermeasures against the intentional introduction or natural occurrence of catastrophic animal, plant, and zoonotic diseases. The Secretary of Homeland Security will coordinate these activities. This effort will include countermeasure research and development of new methods for detection, prevention technologies, agent characterization, and dose response relationships for high-consequence agents in the food and the water supply.

Thus, the suggested ways of fighting an attack are monitoring, warning, cleanup, and treatment.

Vulnerabilities

The forms that an attack on the drinking water supply system could take are as many and varied as are the components of the system that are vulnerable. This chapter will deal with these various methods of attack.

The provision of drinking water to our homes is a complex process that involves many steps, all of which are to some degree vulnerable to compromise by terrorist acts. These envisioned attacks could take one of two forms: a denial of service or a deliberate attempt to contaminate the water. A denial-of-service attack is when the supply of water being provided to the end user is interrupted by tampering or destruction of one of the components in the delivery system. A water contamination event is just what the name implies. The finished water delivered to the end user is contaminated. This contamination can take the form of toxicants or biological agents that cause death or illness in either the short or the long term, or they can be nuisance contaminants that affect the taste or odor of the delivered water in such a way as to make it unusable, resulting in an effective denial of service outcome.

Figure 4–2 illustrates the various components of the drinking water supply network that could come under terrorist assault. An attack could occur anywhere in the system by a variety of modes. The following sections in this chapter will detail some of these attack scenarios and indicate which are the more likely to occur.

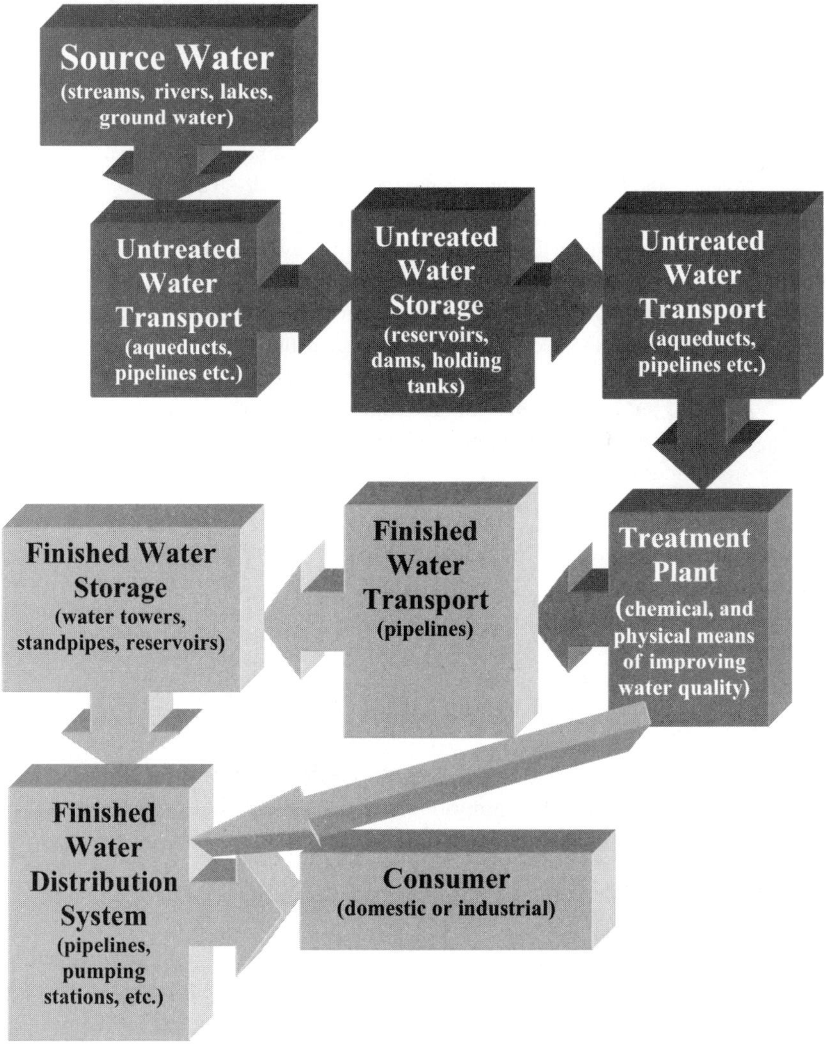

Fig. 4–2. The systems that provide our water are complex and offer a variety of possible points of attack. While some components of the system are more vulnerable than others, none is completely secure.

Source water

The sources of our drinking water are various. Each municipality has its own source or sources. These can include rivers, lakes, streams, groundwater from deep wells, groundwater from wells affected by surface water, desalinization of ocean water, or even recycled sewage water. Some places rely upon a combination of sources—for example, a river in addition to groundwater wells. Importantly, each water provider's source is different, and the characteristics of those sources also differ.

Source waters are, for the most part, immune to physical attacks that would put them out of commission. Theoretically, a group could construct a dam upstream of a drinking water source and prevent the flow to those downstream who use that source, but for terrorist organizations, this is exceedingly impractical. It does, though, have some validity when one country has control of the sources of water flowing into and providing drinking water for another country. However, this is a departure from the realm of terrorism and enters that of international conflict; this action has actually been threatened in a recent dispute between Singapore and Malaysia.[3]

While source waters are not vulnerable to physical attack, they are vulnerable to contamination. As a general rule, source waters are easy to access. It would be impossible to guard all of the points along a river that a terrorist could use as a gateway to the water. Large lakes and rivers are often used for dual purposes, including recreation and transportation, where public access is freely permitted. How vulnerable a specific source is to contamination is directly related to its size and type (fig. 4–3).

Fig. 4–3. The relative vulnerability of a water source has a lot to do with its volume. Large sources, such as a lake, may be difficult to contaminate, whereas small streams may be more susceptible to contamination.

Groundwater is fairly immune to intentional contamination, although examples of industrial pollution have shown that it does occur. Also, a groundwater source could be accessed and a contaminant could be introduced through the wellhead. Groundwater contamination is notoriously difficult to clean up and could result in the source becoming permanently unusable.

Large lakes and rivers, while easier to access than groundwater, have an advantage in terms of protectability owing to the shear volume of water involved. In the water industry, it is said that "the solution to pollution is dilution." Large sources are much less vulnerable than small sources. Also, because many of the potential contaminants are degraded over time by exposure to sunlight and the elements, these sources have a self-cleaning characteristic that makes them difficult, though not impossible, to contaminate.

Initially after the attacks of 9/11, government experts declared that, because of this dilution factor, our water supply systems were fairly secure. Ronald Dick, deputy assistant director for the FBI's counterterrorism division, stated in testimony before congress, "In reality, targeting the water supply may prove difficult. In order to be successful, a terrorist would have to have large amounts of agent."[4] EPA administrator Christie Whitman stated on October 18, 2001,

> People are worried that a small amount of some chemical or biological agent—a few drops for instance—could result in significant threats to the health of large numbers of people. I want to assure people—that scenario can't happen. It would take large amounts to threaten the safety of a city water system. We believe it would be very difficult for anyone to introduce the quantities needed to contaminate an entire system.[5]

As will become apparent throughout the course of this chapter, these experts' opinions as to the safety of our water supplies may be valid in regard to source water but may not hold true for other components of the system.

Essentially, the amount of a compound that would be needed in order to contaminate a source water supply is dependent on the toxicity of the compound and the volume of the supply. Source waters are commonly measured in acre-feet. This is the amount of water that it would take to cover one acre to a depth of one foot. Each acre-foot contains 325,851 gallons. Source water supplies commonly fall in size range of thousands of acre-feet.

Table 4–1 shows how much it would take of a few toxic compounds to contaminate a very small, a small, and a medium-sized source to the LD_{50} level. The LD_{50} is defined as the amount that would kill 50% of the adult males (assuming a weight of 70 kg) if they consumed one liter of the water.

Table 4–1. Weight of material that it would take to contaminate a reservoir to the level that 50% of the adults males (70 kg) drinking one liter of the water would be expected to die.

Toxin	10 Acre-feet Very Small	100 Acre-feet Small	1000 Acre-feet Medium
Sodium cyanide	7,614 lbs	76,142 lbs	761,421 lbs
Sodium fluoroacetate	3,807 lbs	38,071 lbs	380,710 lbs
Strychnine	1,827 lbs	18,274 lbs	182,741 lbs
Nicotine	1,686 lbs	16,859 lbs	168,596 lbs
Endothall	1,577 lbs	15,772 lbs	157,723 lbs
Ricin	57 lbs	570 lbs	5,700 lbs
VX nerve agent	29 lbs	293 lbs	2,931 lbs
Botulinum toxin	15 lbs	152 lbs	1,523 lbs

As should be plain from a review of the data in table 4–1, contaminating a source water to the extent that it could cause mass casualties except in the case of the most toxic substances and the very smallest of sources becomes logistically impractical. The shear volume of chemicals involved is staggering and would require a convoy of dump trucks full of chemicals for even a medium-sized reservoir. This also fails to take into account the environmental degradation of the chemicals or that most of these compounds are susceptible to removal or to degradation by chlorination in the treatment plant. If more were present than could be degraded by normal chlorination, this would be quickly detected at the treatment plant. Therefore, in regard to source water, the government statements issued immediately after 9/11 could be considered correct—but only when the site of contamination is a very large body of water.

While the infliction of mass casualties through an attack on source water is unlikely, an attack making the water unpalatable for domestic use is much more feasible. Doing so would require only a small amount of a compound such as methyl *tert*-butyl ether (MTBE), a common gasoline additive that has a taste threshold of 39 ppb. This means that it would take only a little over 100 pounds of MTBE to contaminate a thousand-acre-foot source water supply such that its taste and odor were affected. Attacks such as this, however, do not fit the goals and aspirations of international terrorism and would more likely occur by accident, be orchestrated by a disgruntled employee or an annoyed customer, or result from simple vandalism.

Therefore, while source water in not impossible to attack, the chances of a mass casualty event involving source water are very low. Incidents involving source water are much more likely to involve accidents or nuisance-type attacks.

Untreated water storage

Untreated water storage includes dams, reservoirs, and holding tanks. They have many of the same attributes as source waters. Like source water supplies, they also vary dramatically in size, and their vulnerability to attack is directly related to size. They tend to be more closely watched than source waters and are not as large geographically as sources; this makes them a little harder to access, but they are not difficult to approach. Many are of a dual-use nature.

Initially after 9/11, much of the emphasis on security in the water sector was directed toward dams and reservoirs. Guards were placed on large reservoirs to prevent attack, and many dams and reservoirs that are also tourist attractions, such as Hoover Dam, were temporarily closed to the public. There was concern that a terrorist could contaminate the water supply behind these dams or cause the dam to fail by physically attacking the structure.

Reservoirs have limited vulnerability to contamination due to their volumes. The same constraints apply to them as to source waters, and a mass-casualty contamination event seems unlikely, although nuisance attacks are still very possible. Unlike source water, however, dams have an added vulnerability: namely, the damage that could be caused if they were subjected to catastrophic failure as a result of a physical attack. In the event of a dam failure, the potential energy of the water stored behind even a small dam can cause loss of life and great property damage if there are people downstream.

According to the Federal Emergency Management Agency (FEMA), there are now more than 10,000 dams in the United States that are classified as having high-hazard potential, meaning that their failure from any means, including a terrorist attack, could result in loss of life, significant property damage, lifeline disruption (critical resources used in everyday life, e.g., power, water, and transport), and environmental damage. The Dam Safety and Security Act of 2002, which was signed into law on December 2, 2002, established the coordination of federal programs and initiatives for dams by FEMA and the transfer of federal best practices in dam security to the state level. The act also includes resources for the development and maintenance of a National Dam Safety Information Network, as well as the development, by the National Dam Safety Review Board, of a strategic plan that establishes goals, priorities, and target dates to improve the safety and security of dams in the United States.

While the potential for disaster exists, is this a valid terrorist threat? (e.g., see fig. 4–4.) Dams are intentionally designed to withstand great pressures. A simple explosion occurring on the surface of a dam is not liable to cause any significant damage. Many larger dams would be unaffected by a surface explosion of any device smaller than a nuclear weapon. However, an explosion designed to detonate under the water near a dam could form a pressure wave large enough to affect the structural integrity of the dam. A large breach is not needed; only a small hole or crack would be necessary for the pressure of the water behind the dam to cause a catastrophic failure.

Fig. 4–4. This small dam (A) *holds back an approximately 8,000-acre-foot reservoir* (B) *directly above a small city of 150,000 people* (C).

This mode of attack was actually exploited by the British air force in World War II. After several unsuccessful conventional bomb attacks on dams, the British instigated Operation Chastise, which entailed the development of specialized bombs designed to sink along the surface of a dam and explode underwater, resulting in a dam-bursting pressure wave. The development of the bombs was quite complicated and took a long time. The goal was to burst several dams and inundate the Ruhr Valley, the center of German wartime industrial production.

After several months of bomb development and practice, Operation Chastise was put into effect successfully with the breaching of several dams. The destruction of the Moehne and Eder dams poured around 330 million tons of water into the western Ruhr region (fig. 4–5). As far away as 50 miles, the destruction was severe. Industrial production was affected because mines and factories were flooded. Roads, railways, and bridges were destroyed. Many deaths resulted from the destruction of towns and houses. In total, 1,294 people were killed, 749 of them Ukrainian prisoners of war from a camp just below the Eder Dam. After the operation, Barnes Wallis, the designer of the dam-busting bombs, wrote, "I feel a blow has been struck at Germany from which she cannot recover for several years."

Fig. 4–5. The Moehne Dam was burst by Allied attacks during World War II. While modern dams are vulnerable to terrorist assaults, the threat is unlikely and easily guarded against.

This turned out not to be the case. Operation Chastise did not have the military effect that was desired. After only a short time, full water output was restored, thanks to an emergency pumping scheme inaugurated only the previous year, and the electricity grid was again producing power at full capacity. The raid proved to be costly to the Allies in lives (more than half the lives lost belonging to allied prisoners of war), but was no more than a minor inconvenience to the Ruhr's industrial output. Nevertheless, the pictures of the broken dams proved to be an immense morale boost to the Allies, especially the British.[6] A similar morale and propaganda boost would no doubt accompany any terrorist activity along these lines.

It took a concerted allied operation to successfully develop a program capable of bursting large dams. Such an operation undertaken by terrorists is unlikely. It would, however, be possible for terrorists to develop a mobile bomb based on a large tanker truck, filled with explosives, that could be driven across a large dam with a roadway on top of it. The driver could then crash the truck into the water alongside the dam. The truck could be rigged to act as a depth charge that would explode along the underwater surface of the dam, thus creating the shock wave needed to burst the dam.

This scenario could be easily prevented by restricting traffic along the tops of the largest and most vulnerable dams. Government officials are aware of this threat, and they are taking it seriously, as evidenced by the following headline from the *New York Daily News* of September 2, 2004: "Bomb study convinces exec to close road atop NY dam." The article details how the county executive, Andrew Spano, was persuaded to leave the road across Kensico Dam, in Westchester County, closed to traffic. Spano said that West Lake Road, which he had planned to reopen on September 4, would remain closed permanently, in the interest of public safety. He had originally planed to reopen the road, but the study changed his mind. New York City, which owns the Kensico Reservoir, behind the dam, had undertaken a bomb blast study to evaluate the dam's vulnerabilities. Christopher Ward, New York City's environment commissioner, was opposed to reopening the road and commissioned the study. According to Ward, if the dam were lost or damaged, "about 250,000 residents could face catastrophic flooding and New York City could lose most of its drinking water."[7]

Untreated water transport

In many cases, the facilities for treating water and the areas where it is needed are quite some distance from the source. It is necessary to transport the water from the source to the treatment plant—by means of aqueducts, pipes, canals, and ditches. These transport conduits can be anywhere from a few hundred yards to hundreds of miles in length. In many instances, they pass through remote areas where little or no physical security exists. Like source waters, they present a tempting and easily accessible target owing to their geographic extent; they too vary in size and volume, and their vulnerability to contamination is directly related to size. Transport systems are not as large as source waters, because they represent the actual supply that is going to the treatment plants for use; therefore, they represent only a fraction of the volume of the source water from which they are derived. Transport conduits do, however, tend to be quite large, and those mechanisms that are left open to the elements, like ditches, do have some of the same characteristics, with regard to natural attenuation and degradation of contaminants, that apply to source water.

Furthermore, the location of transport conduits in the system prior to treatment means that they offer the same barriers to contamination as the source water itself. While not all contaminants would be removed at treatment, it does provide a substantial line of defense. The largest threat to these systems is not contamination—although it is definitely possible—but rather physical disruption of the supply. Remote locations and lack of security make the transport mechanisms susceptible to disruption via conventional explosive attack, especially when they cross valleys via bridges or aqueducts (fig. 4–6). A well-placed explosive could destroy the aqueduct and disrupt the supply of water to the treatment plant. Is this form of attack likely?

Fig. 4–6. The Central Arizona Project Aqueduct is the largest and most expensive aqueduct system ever constructed in the United States. Like many aqueducts, its long length and remote location make it a tempting target; however, any physical damage would most likely be transient in nature. Pumps and pump stations may be an exception.

Physical attacks on source water transport, while a serious concern, would result in few if any casualties; therefore, they do not fit the agenda of most terrorist groups. Also, physical disruption of the water supply in most cases is easily reconciled (in at least a temporary mode) and would not result in lasting damage. One point where this may not hold true is a physical attack on the massive pumps that some of these transport operations rely on to move water up a gradient, which could not be easily replaced if knocked out.

Treatment plants

In most cases, the treatment plant represents the last barrier between natural, accidental, or deliberate contamination and the final end user. In many cases, it is also the final point where continual routine monitoring of the chemical and physical characteristics of the water occurs. Like the previous components of the water supply system, treatment plants can vary greatly in size and in the amount of water treated per day. Some small treatment plants that service small communities may treat only a few thousand gallons of water per day, while those serving large cities may treat tens of millions or even hundreds of millions of gallons per day. The James W. Jardine Water Purification Plant, in Chicago, has a capacity of 1,440 million gallons per day and serves over five million customers (fig. 4–7). Different plants also use different methods of treatment and purification to make the source water potable. The method of treatment varies depending on the source water quality. Some sources require simple settling and disinfection, while others require a more elaborate process.

Fig. 4–7. Large treatment plants such as Chicago's Jardine Water Purification Plant, on the shores of Lake Michigan, can treat over a billion gallons of water per day; plants serving small communities may treat only a few thousand gallons. Either could be a viable terrorist target.

The treatment plant offers the would-be terrorist many opportunities to inflict damage. Al Qaeda and other terrorist organizations are well aware of this and have expressed interest in recruiting workers in such facilities to orchestrate such an attack, as stated in an FBI bulletin issued on August 11, 2004:

> Information recently brought to the attention of DHS and FBI indicates that prior to the September 11, 2001 attacks; terrorists discussed possible attacks against U.S. facilities and systems to disrupt drinking water supplies serving major urban areas, which include large-capacity water reservoirs and water treatment facilities. Although no specific targets were selected, one specific site in the Northeastern United States was mentioned as an example.
>
> While the original thought focused on large capacity water supplies, terrorists thought it would be futile to attempt to directly poison a large water reservoir because of the dilution factor. Rather, they focused on the possibility of poisoning the water during the water treatment process. Terrorists mentioned inserting a poison (not further identified) into the chlorination section of the water treatment facility. To accomplish this objective, they discussed recruiting insiders to work with them. DHS and FBI have no evidence that any operatives were dispatched to the United States after 9/11 to further plan or carry out such attacks. Nevertheless, DHS believes this information should be viewed as one of many legacy plots that could be revisited by terrorists. DHS and FBI further assess that information discussed by the terrorists exhibits a certain degree of operational sophistication and is of particular concern for largely unattended drinking water or wastewater treatment facilities.

Water treatment operations in most industrial countries, including the United States, are a complex, multifaceted process. The number of treatment steps varies from location to location and is generally a function of the source water quality. Very clean or nearly pristine surface waters and groundwaters tend to need less processing than surface waters that have been more impacted ecologically; however, no set procedure exists that can be used in every single case. A variety of disinfectants (chlorine, chlorine dioxide, monochloramine, etc.) are utilized. Some facilities remove metals through a softening process, while others do not; some have to adjust pH, while others do not; and some add fluoride to the finished water, while others do not. Whether the water is prefiltered will vary by system. In addition, the procedures at some facilities incorporate processes such as reverse osmosis, ultraviolet (UV) light, and pulsed UV. The basic operation in most locations follows a set pattern, as illustrated in figure 4–8. While not every plant has every step shown here, the general structure is similar in most cases.

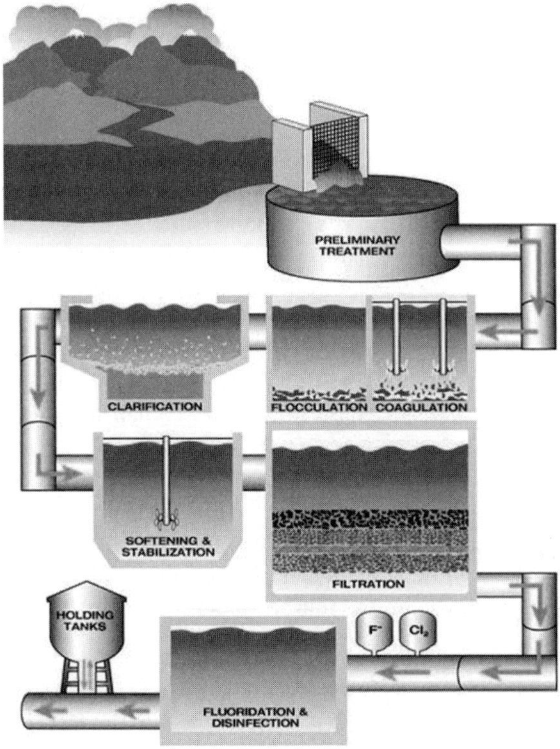

Fig. 4–8. A typical water treatment plant. The intentional addition of chemicals during the treatment process is of especial concern when assessing vulnerabilities. (Diagram courtesy of the Hach Company)

There are many ways that a terrorist could inflict damage once having infiltrated a drinking water treatment plant. Physical assault anywhere in the system could shut down the flow of water. Physical assault could also disable the system, such that the flow of water provided to the customer was uninterrupted but the treatment process was no longer effective in cleaning the water. This is an attack that is easy to mount. One method would be to simply shut down or disable the chemical treatment operations, so that the contaminants that naturally occur in the water are no longer removed. This could be as simple as shutting a valve so that chlorine and other disinfectants are not added, allowing bacteria and other organisms that are usually destroyed in the treatment process to reach the end user. Many people could become ill, and deaths could result. In May 2000, in Walkerton, Ontario, an unintentional failure to properly disinfect the water caused illness in more than 2,000 people and resulted in several deaths.

This form of attack may have dire consequences if orchestrated to coincide with another particular form of attack—namely, an intentional disruption of the process at a wastewater treatment plant that is upstream and feeds effluent into the same source water that the drinking water treatment plant relies on for its raw water. This could dramatically increase the biological load of the source water, thus increasing the chances that it will cause illness.

While the scope of this book is drinking water security and not wastewater security, this is one point where those topics overlap. Most wastewater treatment facilities rely on a process called activated sludge digestion to degrade the organic component of the sewage that they treat. In this biological process, live bacteria are used to digest the incoming material. If the bacteria were to be killed, the digestion would no longer occur. In this case, the overwhelming load of organic material in the incoming sewage would remain undigested. When this undigested material reached the point in the system where the effluent is disinfected prior to being released into the receiving water, the large organic load in the undigested sewage would overwhelm the disinfectant (e.g., chlorine) used in the process. This would result in the dumping of contaminated sewage (essentially raw sewage) into the source water. A large treatment plant could pour millions of gallons of sewage into a receiving body and overwhelm the source water's natural capacity for dilution.

The deactivation of the activated sludge cultures would be quite easy to accomplish, as there would be no way to stop someone from dumping down a drain a variety of chemicals that would be effective in disrupting the system. In fact, this occasionally happens accidentally. The release of raw sewage into source waters is quite common and can be the result of a poisoning and shutdown of the system as described previously; of the loss of power, resulting in the loss of the pump-driven air supply needed for the digestion to occur; or of the overwhelming of the system by a massive flow of water, from storm runoff.

Consequently, millions of gallons of untreated sewage is released to source waters in the United States every year. Normally, however, this is not a severe concern for drinking water, because the treatment process is capable of removing the contaminants and turning it into potable water (fig. 4–9). It becomes a concern only in a coordinated attack scenario (i.e., one that shuts down both processes at the same time). In this case, there is a possibility of contaminated water reaching the end user.

Another important design aspect of drinking water treatment plants is the intentional addition to the water of various chemicals during the treatment process. These can include flocculating agents, caustics, acids, disinfectants, and fluoride among other chemicals. All of the dosing equipment is in place at the facility to feed massive quantities of solid, liquid, or gaseous chemicals into the finished water supply. If terrorists gained access to the plant, they could feed any of a large number of toxic substances into the system. There has also been some concern that terrorists would not even have to gain access to the facility to mount such an attack. It may be much simpler, from the terrorist viewpoint, to infiltrate one of the companies responsible for delivery of chemicals to the treatment plant. They could then replace or contaminate the usual shipment with a toxic compound and deliver it to the plant as normal. Then, when the plant operators added the replaced or adulterated treatment chemical to the water, they would in effect be poisoning the finished water.

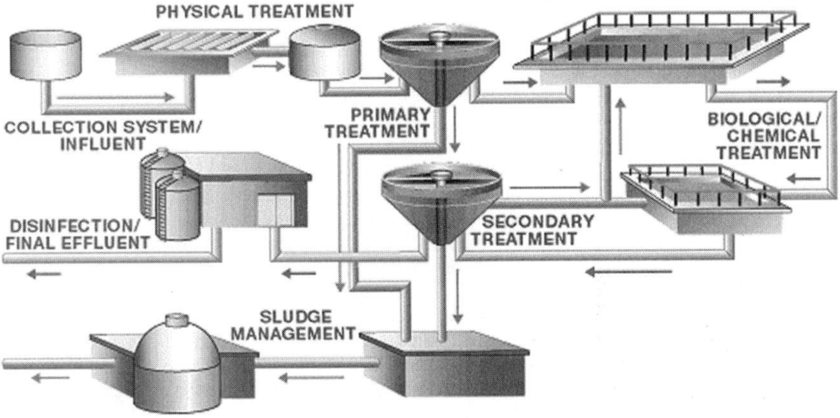

Fig. 4–9. Biological treatment is the key to the effectiveness of most wastewater treatment processes and could be easily disrupted through an intentional attack introducing any of a wide variety of chemicals into drains. (Diagram courtesy of the Hach Company)

While the introduction of toxic chemicals from an outside source is a grave concern, it would not even be necessary to bring the toxic material in from the outside. Many of the normal treatment chemicals already present at the site can be toxic. All that would be needed to cause harm would be for the terrorist to increase the amounts of these usually benign chemicals up to toxic levels.

For example, fluoride is usually fed into the system at a rate that gives a final concentration at the tap of 0.8–1.2 mg/L. The EPA maximum contaminant level for fluoride is 4 mg/L. Fluoride can be extremely toxic at higher doses. This is especially true for young children, in whom the lethal dose can be as low as 16 mg/kg. Patients undergoing dialysis are even more susceptible. There have been past incidents where equipment failure or operator error has resulted in the overfeeding of fluoride into drinking water supplies. Many of these incidents have resulted in illness, and a few deaths have been documented (table 4–2). It would be very easy for a terrorist to simply turn up the release from a feeder to achieve similar results.

Table 4–2. Past incidents of accidental fluoride poisoning

Incident Date	City, State	Details
10/24/03	Marlboro, MA	A valve malfunction caused excess fluoride to enter into the water system, bringing the fluoride level to 24 ppm (6 times the legal maximum)
6/4/2002	Dublin, CA	23 employees of Humphrey Systems Inc. feel ill after using water fountains. Tests at Humphrey Systems on Tuesday showed concentrations of up to 200 milligrams per liter.
7/28/2000	Wakefield, MA	Water system overdosed with fluoride—people were exposed to 23 ppm
11/1993	Middletown, MD	Excess amounts of water fluoridation treatment entered the water supply, raising the level to 70 ppm. Citizens were informed not to drink or cook with the water
8/1993	Poplarville, MS	40 persons poisoned; 15 sought treatment at hospital. Pizza Inn manager was the first to notify city officials after several customers became ill.
7/16/1993	Chicago, IL	Three dialysis patients died and five experienced toxic reactions to the fluoridated water used in the treatment process.
5/1993	Kodiak, AK	Although equipment appeared to be functioning normally, 22–24 ppm of fluoride was found in a sample.
1/1993	Sarnia, ON Canada	Computer-controlled system had failed to shut down.
5/1992	Hooper Bay, AK	Poor equipment and lack of a qualified operator. One death, 260 poisoned; one airlifted to hospital in critical condition.
2/1992	Rice Lake, WI	High winds caused volt lines to connect, causing conductors to burn to ground and a jumper to fail, resulting in failure of the anti-siphoning device, causing fluoride to pour through the pipes. Pump overfed fluoride for two days. Residents vomiting, levels thought to have reached 20 ppm.
1991	Portage, MI	Injector pump failed. Fluoride levels reached 92 ppm and resulted in approx. 40 children developing abdominal pains, sickness, vomiting, and diarrhea.
10/1990	Westby, WI	Equipment malfunctioned, fluoride surged to 150 ppm. The water utility supervisor said he had expected the fluoride to be ten times normal since it had burned his mouth. The fluoride corroded the copper off the pipes in area homes, 70 times higher than the EPA recommended limit.
3/1986	New Haven / N. Branford, CT	The fluoride peaked at 51 ppm. 18% of customers had acute health effects.
10/6/1981	Jonesboro, ME	Equipment had been shut down due to faulty valve controlling the quantity of fluoride going into the drinking water. 57 students, teachers and principal taken to hospital. 38 were administered regurgitants to make them vomit the fluoride, and milk to counteract the poison. Two were admitted to the hospital for several hours for fast heartbeat.
8/10/1981	Potsdam, NY	The diffuser, a plastic pipe that controls flow of fluoride into the water system broke off, allowing the entire contents of a drum of fluoride, ten times a normal "dose," into the water supply. Village residents were without potable water, and in a state of "water emergency" from 2:00 p.m. until 11:00 p.m.
8/30/1980	Vermont Elementary School	A water fluoridator at a local elementary school was accidentally left running, elevating fluoride levels to 1041 ppm (250 times the legal amount), poisoning 22 individuals attending a farmers market hosted at the school.
11/1979	Annapolis, MD	Valve at water plant had been left open all night. One patient died and eight became ill after renal dialysis treatment. The fluoride level was later found to be 35 ppm.
5/1979	Island Falls, ME	Fluoride machine let extra fluoride into water system while motor head was being changed. "The exact water fluoride level was not ascertained although a water sample at a manufacturing plant was greater than 10 ppm." 5 people suffered gastrointestinal illness.
11/17/1978	Los Lunas, NM	Faulty electric relay switch caused concentrated fluoride to be pumped into water system without being diluted with non-fluoridated water. 34 people had acute fluoride poisoning.
11/22/1977	Harbor Springs, MI	Power lines controlling city water system electrical signal lines were down. Approximately 189 lbs. of fluoride was accidentally pumped into the city's water system. Four people experienced nausea or vomiting and weakness.
4/16/1974	Manly, NC	Fluoride feeder pump malfunctioned, causing the fluoride solution to be fed into the water system continuously while water pump not operating. 213 individuals experienced nausea after drinking orange juice mixed with water. 201 students and 7 adults vomited.
6/6/1972	Rome, PA	Blockage of BIF feeder by-pass occurred sending excess fluoride into water system as high as 67 mg./L 150 students attending a school picnic vomited after drinking orange juice made with the water.

As noted previously, the treatment plant is often the last location where routine continuous monitoring of water quality parameters is performed. Drinking water metrics such as pH, turbidity, disinfectant levels, and fluoride are commonly tracked by online continuous-reading monitors at larger plants and by regular grab sampling and laboratory testing at smaller facilities. Other tests (as for bacteria and metals) occur off line and at regular intervals.

In situations in which the same person is responsible for both testing and dosing of treatment chemicals, it would be quite easy to elude detection of an attack. This becomes more difficult if separate personnel are responsible or if the data are automatically collected and downloaded to a third party. While disruption to many parts of the system is feasible, the treatment plant is the first location where the possibility of deliberate contamination, in an attempt to cause mass casualties, becomes realistic, owing to the ease of introducing chemicals and the decreased levels of dilution that are afforded at this point in the water delivery system.

Finished water storage

The storage of finished water (ready for drinking by the consumer) is technically a part of the water distribution system, but for clarity, it will be dealt with here as a separate component of water supply. Once the water has been treated at the plant to the quality level required in order to make it potable, it is often placed in finished water storage facilities. Historically, finished water storage facilities have been designed to equalize water demands, reduce pressure fluctuations in the distribution system, and provide reserves for firefighting, power outages, and other emergencies. The size and the design of the facilities vary from system to system. The main categories of finished water storage facilities are ground storage, which encompasses tanks and reservoirs, and elevated storage, which encompasses water towers and standpipes. Finished water storage does not include facilities that are part of treatment plants, such as clear wells.

Ground storage (tanks or reservoirs) can be belowground, partially belowground, or aboveground in the distribution system; at elevations where gravity does not provide the required system pressure, tanks or reservoirs may be accompanied by pump stations. Ground storage reservoirs can be either covered or uncovered. Although the majority are covered, there are 140 uncovered finished water reservoirs in the United States. Covered reservoirs may have concrete, structural metal, or flexible covers.

The most common types of elevated storage are elevated steel tanks and standpipes. In recent years, elevated tanks supported by a single pedestal have been constructed where aesthetic considerations are an important part of the design process. A standpipe is a tall cylindrical tank normally constructed of steel, although concrete may be used as well. The standpipe functions as a combination of ground and elevated storage.[9]

Like any part of the water supply system, the finished water storage components are vulnerable to physical attack. The design of water towers is such that their structure would be easily damaged by a relatively small explosive device. It would be necessary to destroy the structural integrity of only one spindly support to cause a failure. The physical destruction of water storage, however, is not liable to be a terrorist objective. The inconvenience derived from such an attack could be large, but the resulting loss of life would be relatively small.

History bears this out. On the morning of March 19, 1909, in Parkersburg, West Virginia, two very large water tanks holding a total of over two million gallons of water burst (fig. 4–10). The resulting flood devastated the neighborhoods downhill from the flood. Even with the major destruction caused by the accident, only a few people were injured, and only three died. Terrorists with a bomb that could destroy a water tank would probably achieve more spectacular results by using the same explosive device in a more traditional attack, such as to public transportation (which seems to be a favored target of many terrorist groups). Therefore, it seems unlikely that professional international terrorists would mount such an attack.

Fig. 4–10. In 1909, a major structural failure of two finished water storage tanks in Parkersburg, West Virginia, caused extensive damage but few casualties.

This is not to imply that such attacks are not a danger; rather, their implementation by al Qaeda and similar groups is unlikely. If they are orchestrated, they are more likely to be the work of other types of groups—for example, radical environmental groups. In October 2002, ELF, a self-styled eco-defense group that uses direct action in the form of economic sabotage to stop perceived exploitation and destruction of the natural environment, threatened to destroy two water tanks near Winter Park, Colorado, each holding one million gallons. This type of attack, by destroying the infrastructure, would be consonant with their agenda to stop development that they allege harms the environment. The threat against the water tanks was never realized. It would, however, have met ELF's operational agenda of stopping development without causing loss of human life. This is not the same type of agenda as espoused by those groups wishing to cause mass casualties and would thus preclude physical attacks by al Qaeda–type groups.

A contamination attack on finished water storage, by contrast, is an extremely effective mode of assault by those wishing to inflict mass casualties. Intentional contamination of finished water circumvents many of the problems associated with intentional contamination of source water. The dilution factor is much reduced; the treatment process has already been completed; the time to reach customers is drastically reduced, as is exposure to factors such as sunlight; and, therefore, the time in which the toxicant could be degraded through natural attenuation is decreased. Finished water storage facilities vary greatly in size, from several million to only a few thousand gallons. In large cities with tall buildings, dedicated water tanks provide pressure for individual skyscrapers. It would take a relatively small amount of material to contaminate one of these smaller tanks. As these tanks are often maintained not by the water company but by building maintenance personnel, security can be quite lax, and access can be fairly simple.

In many small communities, the water storage facility is located on public property. For convenience's sake, this is often close to a local school building, because the city already owns land there. Al Qaeda has stated that children are a viable and desirable target. A large percentage of the population, not to mention the vast majority of the children, in these small communities can be found at the school building on any given day (fig. 4–11).

Fig. 4–11. In many small communities, the municipal water tower is located on town or city property, often near target facilities such as the local school.

This presents an ideal target, from a terrorist viewpoint. There is the potential for mass casualties, many of which would be children. The hydraulics of most small systems would entail preferential flow to the buildings closest to the water tower, and the school would probably have the greatest flow rate in the area; a contamination event under these circumstances could be highly effective. Also, the attack would occur in a small rural area where the security is liable to be less restrictive than in larger metropolitan areas. Finally, an attack on such a target would strike a segment of the population that have (mistakenly) perceived themselves to be removed from the terrorist threat.

Security has always been a concern at finished water storage facilities. In the past, it has not always been concerns about the safety of the water and water quality that has driven the attempts to secure these facilities. For decades, water tanks have been a favored canvas for graffiti artists. At the least, most rural water towers sport for part of the year a slogan proclaiming the predominance of that year's graduating high school class. Because of the high cost of graffiti removal and the danger to young people climbing elevated structures, many communities have installed some form of intruder security—traditionally, gates and locks—at their storage facilities and have removed ladders that would facilitate the climbing of these structures. Nevertheless, these precautions have been mostly ineffective, as the graffiti that continues to grace storage tanks and water towers proves. If the high school students find the superficial security measures to be easy to breach, then it is obvious that a dedicated terrorist would have no problems in gaining access. Reliance on locks and gates to stop a dedicated individual who intends to do harm is futile, as figure 4–12 illustrates.

Fig. 4–12. Physical security has always been a problem at finished water storage sites. Generations of teenagers have used water towers and storage tanks as canvasses for graffiti. If locks and chains can't keep the kids with paint out, they are unlikely to thwart terrorists with toxins. (Photograph courtesy of Lorianne DiSabatto at http://www.hoardedordinaries.com)

Still, there are difficulties inherent in mounting an attack on finished water storage. While much of the security precautions taken in the past have proven to be ineffective, there is usually some sort of physical security system that must be bypassed. As utilities become more security conscious, old-fashioned gates and locks are being augmented with closed-circuit television (CCTV) cameras, motion sensors, and remote water monitors. In addition, local police have been made aware of the vulnerability of such facilities and have increased their surveillance of such sites.

Moreover, even if a contaminant were introduced to a water tank, there is a problem with mixing and dissemination. Water tanks have historically had problems with hydraulics and the preferential flow of newer water as compared to older water. That is, the hydraulic mixing of the tank's water may not be adequate to distribute a contaminant evenly. If a compound is simply dumped into a tank, the result may be a large quantity of water with only a small trace of the compound present and a short slug of more concentrated material. This would result in fewer consumers being exposed to a high enough dose to be affected than would be expected. While finished water storage is a definite target, the real vulnerability is in the finished water transport or distribution system.

Finished water transport—the distribution system

The distribution system delivers drinking water primarily through a network of underground pipes to homes, businesses, and other customers. While the distribution systems of small drinking water utilities may be relatively simple, large systems serving major metropolitan areas can be extremely complex. One system, for example, measures water use through 670,000 metered service connections and distributes treated water through nearly 7,100 miles of water mains that range from 2 inches to 10 feet in diameter. The systems also contain pumping stations, disinfectant booster locations, storage tanks, and fire hydrants.[10] Using physical means to attack such a widespread and diverse infrastructure would make little operational sense; however, a terrorist attack causing casualties through contamination is a credible scenario.

When the U.S. government declared, after 9/11, that the water supply was safe owing to the dilution factor, they failed to address the vulnerability of the distribution system. The concept that dilution provided security for the water supply system was short lived. It wasn't long before government officials and industry experts realized that the crucial vulnerability to contamination was in the distribution system.

In October 2003, the Government Accountability Office (GAO) reported to the Senate that the distribution system was the area most vulnerable to attack. John Stephenson, the Director of the Natural Resources and Environment Division of the GAO stated, in testimony before congress,

> Nearly 75% if the experts identified the distribution system as the most vulnerable of all system components. In a typical drinking water system made up of a supply source, source water facility and distribution system—the distribution system was cited as the greatest vulnerability because it is easily accessible at so many points, such as a fire hydrant or standpipe within a building. In fact the water is post treatment, meaning that a chemical, biological or radiological agent would be virtually undetectable until it was too late to prevent harm.

Many noted industry experts view water distribution as the most susceptible to terrorist attacks. John Sullivan, chief engineer for the Boston Water and Sewer Commission and the president of the Association of Metropolitan Water Agencies, concluded that water distribution systems remain the most vulnerable to terrorist threats and could spread highly concentrated amounts of poison to a few thousand homes or businesses. The chair of the National Academy of Sciences's Water Science Technology Board found water distribution systems to be difficult to secure and recognized that, while such systems may affect a smaller population than some other water system components, exposure of a entire city is not needed if the terrorists' goal is fear and anxiety.

In the words of Mr. Stephenson, of the GAO,

> Source water, such as a reservoir, is pretreatment, and is not a very effective way to contaminate drinking water. However, when is the last time you saw a truck backed up to a fire hydrant and assumed he was taking water out of the system? He could just as easily be putting a contaminant into the system. Since it is post treatment, it goes directly to the homes and businesses from there. And that is why I think our experts cited that as a high vulnerability.[11]

As far as contamination is concerned, the consensus expert opinion is that if it is going to occur, then it is most likely to happen in the final distribution system. On the one hand, increased emphasis on physical security measures like gates and fences at water treatment plants has to a certain extent made these hardened targets. The distribution system, on the other hand, is by its very nature not capable of protection through such measures. This effectively makes the distribution system a soft target. Terror groups have long altered operational imperatives to take advantage of soft targets when initial targets are hardened.

Even once the likelihood that an attack with CBR agents would take place somewhere in the distribution system is acknowledged, several misconceptions about this type of attack persist. Historical dogma (incorrectly) holds that such attacks require the assistance of several technicians, are expensive to carry out, and require complicated and expensive pumping equipment to inject contaminants into a pressurized system. Further incorrect assumptions include that such attacks would be very limited in the area of exposure and that only a single building or neighborhood would be affected.

The government agencies that are responsible for water security are beginning to recognize that these assumptions are invalid. They have not, however, completely acknowledged the extent of the problem and still cling to some false assumptions, as illustrated by the following transcript of a hearing on homeland security research and development at the EPA ("Taking Stock and Looking Ahead"), which took place before the House of Representatives Subcommittee on Environment, Technology, and Standards on May 19, 2004. The exchange is between Dr. Gregory B. Baecher, Professor and Chairman, Department of Civil and Environmental Engineering, University of Maryland and Mark Udall, Congressman from Colorado; Dr. Paul Gilman, Science Advisor to the EPA and Assistant Administrator for Research and Development, was also present.

> MR. UDALL. With that, Dr. Baecher, if I could direct a question to you, and maybe Dr. Gilman has a comment as well. You note the lack of information in the Action Plan regarding threats to the Nation's wastewater infrastructure. Could you elaborate on the potential harmful effects of attacks to the wastewater systems, and how you would prioritize this research need, and then, if I could add another comment. In looking over some correspondence between the—Hach and the EPA, there is this mention of backflow attacks on drinking water distribution. I am reminded that I have a back pressure valve on my sprinkler system in my home, which is to prevent water backing up into the house. Is this a similar kind of dynamic that is being alluded to here with a backflow attack? I have heard a lot of talk concerning my local water system. If somebody goes to the reservoir and pours some chemical, or an agent in the water, it will be diluted pretty quickly, and then it has to run through the systems that are in place to deliver safe water to homes and to businesses. Is this a way to get around that problem that an attacker would face?
>
> DR. BAECHER. Perhaps I should start with that question first.
>
> MR. UDALL. Yes.
>
> DR. BAECHER. If I may. If you look at the water system, we have water collection areas and reservoirs, which hold that water, and then it is transferred over relatively long distances to a treatment plant, where it is sometimes filtered, sometimes not, and chlorinated and otherwise made suitable for potable water. Then it is put into a distribution system, a pipe network, and distributed to retail users, to people in their homes and to businesses and that sort of thing. If you introduce contaminants at the supply point in the reservoirs, there is quite a lot of dilution. That is not to say that there are no opportunities, but there is quite a lot of dilution, and that water is also subsequently treated, perhaps filtered, perhaps chlorinated. If you go downstream of the treatment plant, though, into the pipe network, in most urban areas, any fire hydrant, any faucet, can be back-pressured to introduce contaminants back into

the water downstream of the treatment. Now, there still are residuals; chlorine and other chemicals in the water to protect it, but nonetheless, you could, on a local basis, within a small, perhaps multi-block area, have significant impact by back-pressuring contaminants at that point, downstream of the filtering and chemical treatment. And I believe that is what you are referring to. There are protective devices that are on the scale that need to be used, which would not be inexpensive, but there are back-pressure devices that can be used to prohibit people from doing that. They typically are not installed in, for example, fire hydrants.

MR. UDALL. Would it take some specialized equipment to actually perpetrate that kind of an attack?

DR. BAECHER. It would not, sir.

MR. UDALL. It would not?

DR. BAECHER. No, just—as long as you can get the pressure sufficiently higher than the pressure in the distribution system, you can basically pump water upstream, if you will.

MR. UDALL. And you could do that in the dark of the night?

DR. BAECHER. Well, you could rent a house on Capitol Hill.

MR. UDALL. Do you have any good news here?

DR. BAECHER. The spatial distribution of the impact would be limited, of course—

MR. UDALL. Yes.

DR. BAECHER. That is some good news.

MR. UDALL. So you wouldn't even have to be doing it in the public domain?

DR. BAECHER. It would not.

MR. UDALL. It could be undertaken in the privacy of your own rented home.

DR. BAECHER. That is right, but I mean, the number of people that would be affected by that sort of attack is relatively limited.

MR. UDALL. Yes.

DR. BAECHER. Because the materials have to move through the distribution network. They won't move that far, and depending on where you attack the distribution system, there may not be that much downstream of it.

While their testimony recognizes the fallacy of the accepted dogma and acknowledges the threat to the water distribution system, not all of the old assumptions about the threat have disappeared: there is still the notion that this is a limited threat. All of these assumptions are false. More recent studies by the Army Corps of Engineers, the U.S. Air Force, Colorado State University, and Hach HST, among others, have shown that CBR attacks could be carried out for five cents or less per lethal dose, that a single individual can obtain or produce effective contaminants in quantity, and that contaminants can be introduced into the distribution system with the aid of inexpensive and easy to obtain pumping equipment via a method called backflow attack. Contamination could be spread across a wide geographic area if proper site location and pumping methods are chosen, resulting in widespread mass casualties.

What is a backflow attack? A backflow attack occurs when a pump is used to overcome the pressure gradient that is present in the distribution system's pipes. This is usually around 80 pounds per square inch (psi) and can be easily achieved using pumps available for rent or purchase at most home-improvement stores or through Internet sources. After the pressure gradient in the system has been overcome and a contaminant has been introduced, Bernoulli effects act as a siphon, pulling the contaminant into the flowing system. Once the contaminant is in the pipes, the normal movement of water in the system disseminates the contaminant throughout the network, affecting areas surrounding the point of introduction, which can be anywhere in the system (e.g., a fire hydrant, a commercial building, or a residence; see fig. 4–13).

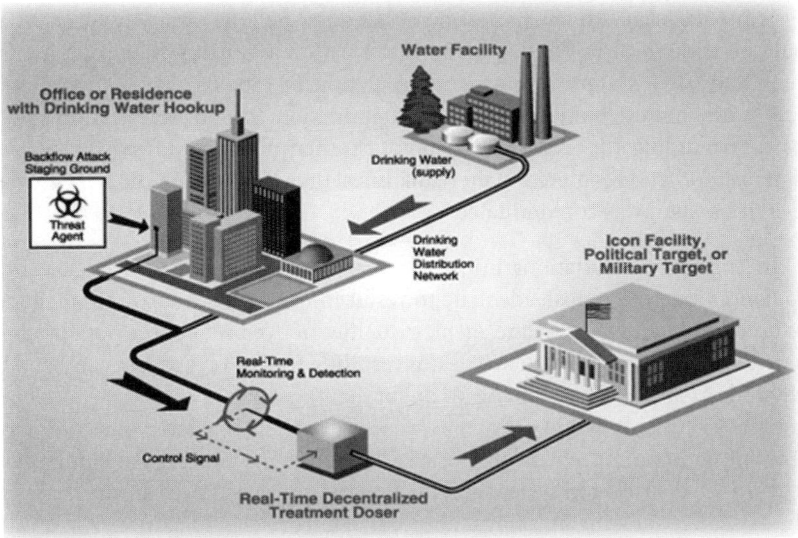

Fig. 4–13. All distribution systems are vulnerable to a backflow attack. (Diagram Courtesy of Hach HST)

Backflow scenarios are not merely conjecture. Backflows occur by accident on a regular basis and are of great concern to the water industry. An organization called the American Backflow Prevention Association is dedicated to the study and prevention of backflow events. Accidental backflow events have been responsible for many incidents of waterborne illness and even death in the United States. According to the EPA, backflow events caused 57 disease outbreaks and 9,734 cases of waterborne disease between 1981 and 1998.[12]

To prevent accidental backflows, many systems have been equipped with backflow prevention devices. These devices function by exploiting an air gap, which either eliminates a cross-connection or provides a barrier. The basic mechanism for preventing backflow is the use of a mechanical device to serve as a physical barrier. The principal types of mechanical backflow preventers are the reduced-pressure principle assembly, the pressure vacuum breaker assembly, and the double check valve assembly; a secondary type of mechanical backflow preventer is the residential dual check valve.

These means of preventing backflow are very useful in preventing the all-too-common accidental event. Note, however, that these devices are installed to prevent *accidental* backflows. They are all physical devices that can be removed or disabled quite easily by a would-be terrorist, thus rendering them ineffective in preventing deliberate attempts at contamination by all but the most amateurish perpetrators.

Intentional dissemination of contaminants through a backflow event represents a critical vulnerability. Studies conducted by the U.S. Air Force and Colorado State University have shown this to be a highly effective means of contaminating a system.[13] In brief, studies have shown that a few gallons of highly toxic material would be enough, if injected at a strategic location and using the proper method, to contaminate a system supplying a population of 150,000 people in a matter of a few hours. Material and significant contamination was not relegated to only the areas surrounding the introduction point. Rather, material flowed through each neighborhood and reentered main trunk lines, thus permeating the next area until the entire system was contaminated.

In computer simulations utilizing a warfare-type material, over 20% of the population received a dose adequate to result in death; when a common chemical was used in place of the warfare agent, casualties of over 10% of the population still resulted. This shows thousands of deaths resulting from a very inexpensive and low-tech mode of attack. There can be no doubt that this form of assault meets all of the terrorists' criteria. It would cause mass casualties, be inexpensive, and offer a good chance of avoiding apprehension. Because little or no monitoring for contamination occurs in the distribution system and the first indications of such an attack are likely to be casualties showing up at local hospitals, a terrorist could launch such an attack and be on a plane out of the country before the first casualty is reported. This is the mode of attack that the would-be terrorists in Rome, as discussed in chapter 3, were attempting when they were disrupted in 2002.

These sorts of attacks can be launched from any access point to the water system. Wherever water can be drawn out, material can be forced back into the system. Some areas, however, are more vulnerable than others. Access points near high-flow areas and larger pipes would be favorable, because they would disseminate the material to a wider area more quickly; however, any access point other than at the end of long deadhead lines could be used to effectively access the system.

What materials could be used to contaminate the system? On the basis of the large number of accidental backflows that occur and the penchant that terrorist organizations have shown for attacking water, the distribution system is a prime candidate for attack. The question remains, are there any contaminating substances available that would be effective as the toxic instrument in such an attack? A bona fide terrorist is virtually inundated by possible candidate substances that would be very effective. The possibilities are virtually endless, and that is in part why protecting against and/or detecting such an attack is so difficult.

Which of the myriad possible agents would most likely be deployed in a terrorist assault is a matter of conjecture. The possible number of chemical and biological substances that could be used is very large.[14] A variety of lists specifying likely agents are publicly available in the current literature, such as the lists compiled by the Centers for Disease Control and Prevention (CDC) and the military (see app. A). There are also lists that have been compiled but are unavailable, owing to security concerns, such as the list compiled by the EPA. Many of these lists are similar; however, no two lists are identical, and in several cases, they are contradictory. In practice, some of the lists, even those that were composed specifically for water, don't make a lot of sense. They fail to take into account factors other than simple toxicity, such as solubility, stability in water, degradation by chlorine, availability to a terrorist, ease of handling, danger to the terrorist, and detectability by the consumer, among others.

Even with all of these criteria in mind, there are still a plethora of compounds from which to choose. While an in-depth listing of the best candidates has been prepared, it is not presented here, for security reasons: the last thing anyone wants to do is provide a road map to aspiring terrorists. In the next section, I will, however, discuss broad categories of toxicants and some of their attributes. All of the information presented here is readily available from public sources and thus does not in any way represent a breach of security.

Toxicants Usable in a Water Attack

Heavy metals

According to one definition, the heavy metals are a group of elements between copper and lead on the periodic table of elements, having atomic weights between 63.546 and 200.590 and specific gravities greater than 4.0. They include metals such as lead, cadmium, arsenic, mercury, and thallium. Many of the metals are toxic if consumed in doses larger than the body is able to sequester or secrete. This toxicity is what makes them agents of concern from the standpoint of terrorism readiness. The salts of most of these metals are readily soluble—many in the 100,000 mg/L range. Their toxicity varies greatly, depending on the metal and the specific salt form. For example, the LD_{50} for lead nitrate is 49,980 mg, while for thallium chloride, it is 1,260 mg. Generally, these heavy metals are not the most toxic materials available to a terrorist in terms of a lethality; however; they may have adverse health effects at levels well below their respective LD_{50}s.

As a general rule, salts of heavy metals are odorless and tasteless, although some do have a distinct taste. Although they are readily available from a variety of industrial sources, they are not easily obtained in huge amounts. Organometallic complexes of heavy metals, such as dimethyl mercury, tend to be even more toxic than their inorganic counterparts. Compounds such as dimethyl mercury pose a double threat, because they are also readily absorbed through the skin. This means that exposure could come from a shower or other skin contact. They are also very volatile, so that inhalation becomes a threat. Notably, these compounds pose a risk to people who think they are safe because they drink only bottled water. (Bottled water will be discussed in more detail later.) Organometallic complexes are, however, more difficult than the inorganic salts to obtain in quantity. Overall, heavy metals are a valid terrorist threat.

Herbicides

A herbicide is a pesticide used to kill unwanted plants. Selective herbicides kill certain targets while leaving the desired crop relatively unharmed; these are often derived from plant hormones and act by interfering with the growth of the weed. Herbicides used to clear waste ground, by contrast, are nonselective and kill every plant with which they come into contact. Because they are specifically designed to be toxic to plants and not animals, as a general class, herbicides tend to be less detrimental to human health than some other compounds; there are some notable exceptions. On the one hand, glyphosate, commonly known as Round-up, has an LD_{50} of approximately 3,412,110 mg; this is not exactly toxic. On the other-hand, endothal, a common aquatic herbicide, has an LD_{50} of 4,060 mg.

What makes herbicides agents of concern, however, is not so much toxicity as availability. Huge quantities of these chemicals are produced and used every year by the agricultural market. Agricultural chemicals are notoriously easy to obtain in large quantities. Anhydrous ammonia is a common component used in the illegal synthesis of the street drug methamphetamine (also called crystal meth). For years, law enforcement officials have waged a battle to keep this material from finding its way into the illegal drug laboratories. So far, they have been unsuccessful. Every year, the number and volume of thefts seems to increase, even as more security is put into place around anhydrous ammonia tanks.

Herbicides are relatively inexpensive, and because they are not extremely toxic, they are offered for sale on the consumer market. You can buy them at any lawn-and-garden center or home-improvement store. While large quantities would be required in order to cause a mass-casualty event, it wouldn't take much to affect the waters such as to require costly cleanup. Even if few fatalities resulted, the panic caused by the introduction of herbicide-type compounds into a water system could be severe.

Insecticides

Normally, insecticides tend to be more toxic to humans than herbicides. There are a number of different types of insecticides, including chlorinated compounds, carbamates, organophosphates, and plant-derived insecticides. Certain widely used insecticides, owing to their ubiquitous nature, have been placed on threat-agent matrices; however, they are not really serious threats to water. For example, lindane, which was at one time the most widely used insecticide in the world and is found on many lists of threat agents, is limited in water by its insolubility. However, some quite toxic insecticides are very soluble. For example, methomyl has a solubility of 172,000 mg/L, and nicotine, a plant-derived insecticide, is very soluble and very toxic.

Many insecticidal compounds, especially organophosphates and carbamates, have the same mode of action as the military nerve agents. They are cholinesterase inhibitors (or anticholinesterases), chemicals that inhibit a cholinesterase enzyme from breaking down the neurotransmitter acetylcholine, thus increasing both the level and duration of action of acetylcholine. Cholinesterase inhibition is associated with a variety of acute symptoms, such as nausea, vomiting, blurred vision, stomach cramps, and rapid heart rate and eventually death.

Like herbicides, insecticides are also widely used in agriculture and the home market and can be obtained in large quantities with little difficulty. Like the organometallic compound described previously, many of these compounds, such as nicotine, are toxic by skin contact and are volatile; thus, they could cause casualties without being ingested. Some of the most dangerous terrorist threat agents to water are of the insecticide class.

Nematocides

Nematocides are designed to kill nematodes in the soil and have many of the same characteristics as insecticides. Despite a few exceptions, they are more soluble than insecticides, allowing greater penetration into the soil. An example is oxamyl, with a solubility of 280,000 mg/L. Many of these compounds have uses specific to certain types of crops and geographic areas. As agricultural chemicals, they may be obtained in large amounts by a potential terrorist. Many of these compounds are cholinesterase inhibitors with structural similarities to nerve agents (e.g., see fig. 4–14). Their high potential for lethality, combined with their ready availability and their solubility, makes them agents of concern. Like insecticides, nematocides are a definite threat to water supplies.

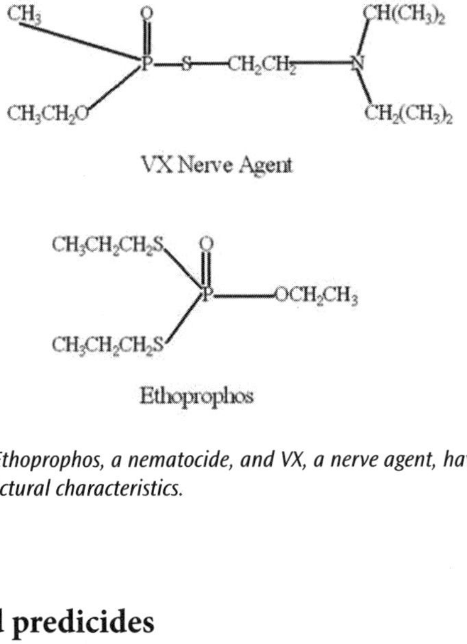

Fig. 4–14. Ethoprophos, a nematocide, and VX, a nerve agent, have many similar structural characteristics.

Rodenticides and predicides

Rodenticides are designed to kill rodents, and predicides are designed to kill predators such as coyotes. While these are technically agricultural chemicals, they are not used nor are they available in the same volumes as other agricultural chemicals. The very fact that these chemicals were designed to kill higher life-forms indicates that they would be toxic to people as well. These chemicals are diverse and work via a variety of modes.

Compound 1080, or sodium fluoroacetate, is one of the most troubling potential threats in this class. Compound 1080 is designed to kill coyotes and other predators. It was used in the United States for many years to poison the carcasses of animals on which scavengers would feed. It is very toxic (LD_{50} of 140 mg), extremely soluble, and virtually odorless and tasteless. Once exposure has occurred, there is no known antidote. It is currently banned for private use in the United States. It is legal only in some states when used in a special collar that U.S. government trappers use to protect domestic sheep. Coyotes attacking the sheep puncture the collar and thereby contact the poison, which kills them. Private stockpiles may still be found in remote agricultural areas. Because of its efficacy in quelling predators of livestock, there is a black market for this material among ranchers. It is also still in widespread use in several foreign counties, for example, New Zealand.

Compound 1080 is considered to be particularly viable as a terrorist threat. In an October 2004 letter, Representative Peter DeFazio, a member of the Select Committee on Homeland Security, took the highly unusual action of urging DHS Secretary Tom Ridge to act immediately to halt manufacture and use of the poison. The DHS took the letter seriously enough to initiate a formal study.[15] These are extremely dangerous threat agents that could cause horrendous casualties if used by the terrorists.

Radionuclides

A radionuclide is an atom with an unstable nucleus. The radionuclide undergoes radioactive decay by gamma ray and/or alpha or beta subatomic particle emission. Radionuclides may occur naturally, but can also be artificially produced. The use of radionuclides as a terror weapon is a distinct possibility.

Terrorists have shown increasing interest in obtaining nuclear material over the past several years. Obtaining high-purity, highly radioactive material, such as plutonium or uranium-238, is difficult. Thus, a terrorist organization that had obtained these materials would not likely be inclined to use them in an attack on a water system. Rather, they would be expected to use them in the manufacture or threatened manufacture of a thermonuclear weapon.

More likely is the use of low-level radioactive material or waste. These materials are not nearly as difficult to obtain. There have been several incidents of attempted trafficking of strontium-90 and cesium-137. A lot of medical waste is classified as radioactive and contains various isotopes of a variety of radiological material. Most household smoke detectors contain the radionuclide americium. While these materials are radioactive, their acute toxicity is a function more of their action as heavy metals than of their radioactive properties. It would take quite a large volume of these difficult-to-obtain materials to instigate a mass-casualty event through water contamination. However, because of the horror and fear instilled in most people at the thought of being exposed to radiation of any sort, even if casualties were low, the psychological aspect of a radiological threat could be severe. This makes radionuclides a possible, even though unlikely, threat to water.

Street drugs

Illegal drugs are not widely recognized as a potential threat agent for a water attack, although LSD has shown up on a few threat agent lists. LSD, gamma hydroxybutyrate (GHB), phencyclidine (PCP), and heroin, among other street drugs, are potential agents that could be employed in a water attack.

Some drugs, such as LSD, exhibit human toxicity in doses that are only matched by some of the militarized warfare agents. While the lethal dose for LSD has been variously reported to be around 14 mg, the dose that causes hallucinogenic effects may be as low as around 140 μg (100 times less), with some effects starting at doses as low as 25 μg. At this dose, no one is likely to die from toxic effects, but extreme collateral damage—from car wrecks and accidents, as well as panic induced by unexpected hallucinogenic experiences—could result. LSD is not easily synthesized in the pure form, which is favored by drug users to avoid causing a bad trip. However, a crude form is easily synthesized in the home laboratory, and it is doubtful that terror-minded chemists are overly concerned with their victims' enjoyment during the attack; moreover, a bad trip more closely fits their intended purposes.[16]

Other drugs, such as heroin, are readily available despite the best efforts of the Drug Enforcement Administration (DEA). Although cost prohibitive to individuals working alone, supplies do exist for well-funded and organized groups. Notably, a large portion of the illegal opiates (e.g., heroin) entering the United States come from Afghanistan and West Asia—that is, from areas where Islamic terrorist cells are at their most active. In 2002, then-director of the DEA Asa Hutchinson said, "The DEA has received multi-source information that Osama bin Laden himself has been involved in the financing and facilitation of heroin-trafficking activities."[17]

Islamic fundamentalist groups are also vehemently opposed to the recreational use of illegal drugs, and their rampant use in the United States is one of the factors they single out when expounding on the moral decay of the Godless West and the superiority of their vaunted Caliphate. It may intrigue the terrorists to use a symbol of Western moral depravity against us. This fits with the al Qaeda attack criterium that, ideally, the attack should use the attributes of a capitalist society and the West against itself.

An attack of this sort has already been attempted. In 2002, Osama bin Laden tried to buy a massive amount of cocaine, spike it with poison, and sell it in the United States, hoping to kill thousands of Americans. The plot failed when the Colombian drug lords bin Laden approached decided to back out of the deal. They determined that, in the long term, it would destroy their very lucrative business, and they also feared massive retaliation from the United States. It has been asserted that bin Laden personally met with leaders of a Colombian drug cartel in 2002 to negotiate the purchase of tons of cocaine, saying that he was willing to spend tens of millions of dollars to finance the deal. Bin Laden hoped that the death of large numbers of Americans from poisoned cocaine would lead to widespread terror. "They wanted to kill thousands of people—more than the World Trade Center," according to an article in the *New York Post*.[18] Thus, illegal drugs should be considered as a possible threat.

Warfare agents

Over the years, a wide variety of compounds have been designed, by the militaries of the United States and other countries, specifically to inflict casualties in battlefield situations. Great fear exists that some of these weapons could fall into the hands of terrorists or that they could produce their own or similar materials in clandestine laboratories. This class of compounds is made up of both chemical agents and toxins. This section will deal with the chemical agents; the toxins will be covered separately later in the chapter.

Warfare agents can be designed to kill, as casualty agents, or to disable, as incapacitating agents. The casualty agents are generally the most feared and include blood agents, choking agents, nerve agents, and vesicants. Warfare agents could find their way into water supplies either through deliberate introduction or through a subsidiary contamination resulting from an airborne attack.

Blood agents (table 4–3) include arsine, hydrogen chloride, and cyanide gas. While arsenic, hydrochloric acid, and cyanide themselves are viable candidates for introduction into the water supplies, gases are liable to be a water contaminant only in a subsidiary role, because of the difficulty in handling them and the because they would be better utilized in an airborne attack. It wouldn't make a lot of sense to deploy these agents in an attack scenario for which they were not designed.

Table 4–3. Blood Agents

Arsine
Cyanogen chloride
Hydrogen chloride

Choking agents and vesicants (table 4–4) are the warfare agents that have been most used by the military. These include the mustard and chlorine agents that saw mass usage in World War I. There are large stockpiles of these weapons available, and many states that are known to sponsor terrorist activities currently have access or have had access in the past to these compounds. Stockpiles of these weapons left over from the World Wars, along with used but undetonated devices, are also discovered on a regular basis. Some of these have been located in areas with a terrorist presence, such as the Philippines and elsewhere in South Asia. It is not beyond the realm of possibility that a terrorist organization could obtain these weapons. Like the blood agents, however, they are not likely to be used with water as the primary target. There are better ways to disperse these chemicals, many of which are not stable in water or are readily degraded by exposure to chlorine (although the degradation products are often still toxic).

Table 4–4. Choking agents and vesicants

Chlorine
Diphosgene and Phosgene
PFIB
Ethyldichloroarsine
Mustard agents
Lewisite 1,2,3,4
Methyldichloroarsine
Sesqui mustard
Nitrogen mustards 1,2,3
Mustard-lewisite mix
Mustard T mix

Nerve agents (table 4–5) are the most feared of the chemical weapons. These compounds usually work as cholinesterase inhibitors and are extremely toxic, causing death at very low doses. These agents were designed for aerial usage, so the data for their toxicity by ingestion are not as detailed as for inhalation. Many of these compounds are easily hydrolyzed by water and/or broken down by chlorine. Unfortunately, many of the breakdown products are also toxic; moreover, depending on the conditions in the water when the compounds degrade, the resulting products of some of these compounds may actually be more toxic than the parent chemical. Once again, while the nerve agents are a definite threat to water, they would probably not be deployed in that role.

Table 4–5. Nerve agents

GA (Tabun)
GB (Sarin)
GD (Soman)
GE
GF
VE
VG
VF
VX

Incapacitating agents (table 4–6) include depressants, such as morphine and fentanyl, and psychedelics, such as LSD. Most of these compounds have other sources and uses other than as warfare agents and are covered elsewhere. An exception are agents such as 3-quinuclidinyl benzilate (BZ), a psychosomatic agent created by the military. These compounds were never created in large quantities, and their exact toxicity and reactions with water are not well known. If terrorists used one of these agents, it would most likely be one of the more common compounds in this class that are more readily available.

Table 4–6. Incapacitating agents

Morphine and Derivatives
Agent 15
BZ
LSD
Mescaline
Phencyclidine
Psilocybin
Amphetamine
Cocaine
Dexamphetamine
Methamphetamine

Plant toxins

Plant toxins are simply extracts of plants that act as toxic substances. The plant world is full of species that produce compounds that are toxic, for defense and for other functions beneficial to the plant. This group includes compounds such as poison hemlock and digitalis, as well as cyanide-based compounds. Some of the agents of greatest chemical concern are plant toxins.

One plant toxin that almost everyone is familiar with is ricin, which comes from the castor bean (*Ricinus communis*). It is on many lists as a potential terrorist agent and was at one time in Iraq's chemical weapons arsenal. Ricin is very toxic and has been used as a weapon of assassination. Terrorist organizations have shown great interest in ricin, partially because of the ease of obtaining the material from a simple extract of beans of the castor plant (fig. 4–15). Several seizures of ricin have occurred, and many documents detailing its manufacture and use have been captured. While the solubility of ricin is limited, it may not be noticed, as it would give only a simple light cloudiness to the water, which is not an uncommon characteristic in many locations. Ricin is extremely resistant to degradation by chlorine[19] and should be considered to be a viable threat agent in water.

Other plant toxins, though not as well known as ricin, may prove to be an even greater danger. The rosary pea (*Abrus precatorius*), also called Indian licorice or jequirity bean, contains a poison known as abrin, which is related to ricin (fig. 4–16). Abrin is very similar to ricin but 75–100 times more toxic; thus, far less material would need to be extracted to yield the amount of poison necessary to contaminate a given body of water. It is native to the Indian subcontinent and has found its way into many other areas. It is widely used in jewelry production because of its bright red beadlike seeds. A terrorist could easily obtain the material to make a toxin from these seeds, or they could simply grow their own.

Fig. 4–15. The toxin ricin is derived from the bean of the castor plant through a simple extraction procedure. Waste from the production of castor oil is 3–5% Ricin. (Photograph courtesy of the Agricultural Research Service, United States Department of Agriculture)

Fig. 4–16. The seeds of the jequirity bean contain the toxin abrin, which is far more potent than ricin. (Photograph courtesy of the United States Geological Society)

Even more common plants may also pose a threat. The common jimsonweed (*Datura stramonium*) is native throughout the United States and is also found in Asia, Canada, and the Caribbean. Jimsonweed is a source of the chemicals atropine and scopolamine. These compounds, if ingested in quantity, are potent hallucinogens, as a detachment of British soldiers discovered in 1676, when they were sent to Jamestown, Virginia (hence the name, from Jamestown weed), to put down Bacon's rebellion. The soldiers were secretly drugged with the weed in their salad,

> and some of them ate plentifully of it, the effect of which was a very pleasant comedy, for they turned natural fools upon it for several days: one would blow up a feather in the air; another would dart straws at it with much fury; and another, stark naked, was sitting up in a corner like a monkey, grinning and making mows [grimaces] at them; a fourth would fondly kiss and paw his companions, and sneer in their faces with a countenance more antic than any in a Dutch droll. In this frantic condition they were confined, lest they should, in their folly, destroy themselves—though it was observed that all their actions were full of innocence and good nature. Indeed, they were not very cleanly; for they would have wallowed in their own excrements, if they had not been prevented. A thousand such simple tricks they played, and after 11 days returned themselves again, not remembering anything that had passed.[20]

The recreational usage of jimsonweed seeds as a hallucinogenic substance has been growing throughout the United States, even though there is substantial danger of accidental poisoning with a fatal dose. The seeds are brewed into a tea and taken to stimulate hallucinations (fig. 4–17). The problem is that the seeds vary greatly in the amount of toxins they contain. What may be an adequate dose to achieve the desired effect from one plant may represent a fatal dose from another plant. This results in several deaths every year from overdoses. These plants are found everywhere in the United States, and it would not take much effort to gather large quantities of material from which to extract toxins. A terrorist could easily take advantage of this ready availability to do harm.

Fig. 4–17. A potent toxin can be brewed from the common plant jimsonweed, found throughout North America and Asia.

Biotoxins

Fillet of a fenny snake,
In the cauldron boil & bake;
Eye of newt and toe of frog,
Wool of bat and tongue of dog,
Adder's fork and blind-worm's sting
Lizard's leg & howlet's wing,
For a charm of powerful trouble,
Like a hell broth boil and bubble

—*Macbeth,* Act 4, Scene 1

As Shakespeare was well aware, a wide variety of toxins are produced by bacteria, algae, and higher animals that have been recognized and utilized by humans for eons (table 4–7). The most well known of these is probably botulinum toxin, or botox. Its notoriety has a lot to do with its toxicity. Botox is produced by the bacteria *Clostridium botulinum,* and it is the causative agent of botulism poisoning. It is one of the most potent toxic substances known to man, with an LD_{50} of 0.4 µg. Whenever the subject of poisoning a water supply is discussed, the topic of botox always arises. It would require only a couple of grams to poison a million-gallon water supply to the LD_{50} level.

Table 4–7. Biotoxins

Toxin	Source	LD-50	Danger of causing mass casualties through water
Botox	Clostridium Botulinum (bacteria)	0.0004 mg	High
Microcystins	Cyanobacteria (Algae)	Unknown ID-50 1–10 mg	Medium
Anatoxin	Cyanobacteria (Algae)	14 mg	Medium
Saxitoxin	Dinoflagellates	0.3–1.0 mg	Low
Tetrodotoxin	Puffer fish and snakes	1–2 mg	Low
Ciguatoxin	Dinoflagellates	0.04 mg	Low
Batrachotoxin	Frogs	0.14 mg	Low
SEB	Staphylococcal bacteria	Usually not lethal Incapacitating	Low

Other toxins are produced by a wide variety of animals, fish, and algae. Microcystins and anatoxins are compounds of concern to the water industry regardless of terrorism, because they occur naturally in source water together with algae blooms. Culturing the algae and extracting the toxins could procure quantities of these materials, which do present a substantial threat. While it would appear on the basis of their toxicity that other biotoxins would be an extreme risk to water supplies, that is usually not the case. The majority of the more obscure toxins, such as the batrachotoxin derived from the South American poison dart frog, would be very difficult for anyone to obtain in large quantities.

However, the most toxic of these compounds, the botulinum toxin, would not be extremely difficult to culture in large quantities. Also, with its increasing use in cosmetic surgery as a wrinkle remover, it would be possible for supplies to be diverted. Botox does have one drawback as a water attack agent. It is extremely sensitive to degradation by chlorine. These toxins are 99.7% inactivated by 3 mg/L free available chlorine in 20 minutes and are 84% inactivated by 0.4 mg/L chlorine in 20 minutes.[21] This means that, to be effective, either a dose must be large enough so that adequate amounts of the material remains undegraded or the chlorine must be removed from the water supply with a co-injection of a reducing material, such as sodium thiosulfate or ascorbic acid.

Mycotoxins

Mycotoxins are toxins produced by molds and fungi. The most notable and widely known are the aflatoxins produced by the mold *Aspergillus flavus*. Also included in this category is the military's T-2, a trichothecene toxin isolated from cereal grains infected with *Fusarium*. These compounds are not as toxic as some of the other compounds more readily available for use in water, nor are they very soluble in water. Although they are thus unlikely candidates for use as a waterborne agent, they could nevertheless find their way into water supplies as the result of an aerial attack. Mycotoxins are fairly persistent in water and resist degradation by chlorine.[21]

Industrial chemicals

Industrial chemicals are a huge category of diverse compounds, including pretty much everything not assigned to other categories. A wide variety of toxic industrial compounds (TICs) and toxic industrial materials (TIMs) are used every day in the United States. By rail alone, more than 160,000,000 tons of TICs are shipped across North America every year. That is more than 2,700 pounds per person![22]

Industrial chemicals include acids, bases, solvents, amines, and cyanides. These compounds are used for a variety of legitimate purposes. For example, the listed industrial applications of sodium cyanide include the extraction of gold and silver from ores; electroplating baths for metal plating; use as a fumigant for citrus and other fruit trees, ships, railway cars and warehouses; use as a raw material in manufacturing of hydrocyanic acids and other cyanide-based materials; and the case hardening of steel.[23] These materials are available in huge quantities and are commonly known to be toxic. Thus, they are a threat not only from trained terrorists but also from any deranged employee who wants to do harm and figures out that dumping some cyanide in the water may be a way to cause damage.

Beyond cyanides, there are other industrial-type materials that may cause damage. For example, methanol is a ubiquitous toxic alcohol used as an industrial solvent. It can cause blindness and even death. Death has been reported after ingestion of less than 30 mL. It is toxic through ingestion, inhalation and absorption through the skin. It is also a very small organic molecule and thus is not removed by many advanced treatment systems, such as reverse osmosis. Owing to their availability and toxicity, TICs and TIMs are a definite threat to water.

Consumer products and nuisance compounds

Many of the compounds already discussed are available to the consumer market. These are usually mixtures of the pure chemicals with a large portion of inert ingredients added. As consumers, we buy more than a quarter million different household products that are used in and around the home, for medication, cleaning, cosmetic purposes, exterminating insects, and killing weeds.[25] These items are valuable in the home and for yard maintenance, but misuse, especially when products are used in inappropriate applications or quantities, can cause illness, injury, and even death. Many of these household-use chemicals are extremely toxic, as evidenced by the large number of domestic accidental poisonings. Each year, more than 6,000 people die and an estimated 300,000 suffer disabling illnesses as a result of unintentional poisoning by solid and liquid substances.[24]

Also included in this category are nuisance compounds, which are not extremely toxic but would make a water supply unusable. These are not liable to be used by international terrorists, but they could be used by disgruntled workers or vandals.

Gasoline, diesel fuel, and kerosene are examples of nuisance compounds that are widely available to the public in large quantities. If any of these compounds were dumped into the water supply, they would not be expected to cause any deaths, but because they are difficult to clean up, the water would be unusable for a prolonged period of time. For example, a small spill of diesel fuel into a reservoir near Glasgow, Scotland, in 1997 left large areas of the city without water for several days while the spill was being cleaned up.[25] Consumer compounds are a significant threat to water supplies.

Biological agents

Bacteria. There is a lot of concern about the use of bacterial agents in an attack against water supplies. Bacterial or other biological agents provide several advantages to a terrorist. Biological agents can cause large numbers of casualties with minimal logistical requirements. The terrorist would have the opportunity to escape the scene of the attack long before biological agents cause casualties, due to the incubation periods of the agents. Person-to-person transmission of several agents (notably plague and smallpox) could perpetuate an epidemic.

Biological weapons are easy to produce and can be used to selectively target humans, animals, or plants. They are also very inexpensive.[26] A study was conducted to determine the costs of causing 50% casualties within a square kilometer. The costs of conventional weapons ($2,000), nuclear armaments ($800), and chemical agents ($600) would far surpass the bargain price of biological weapons ($1) to produce 50% casualties (all in 1969 dollars).[27]

Agents can be easily procured from the environment, universities, biological supply houses, and clinical specimens. In 1995, Larry Wayne Harris, a microbiologist with ties to a white-supremacist group, ordered *Yersinia pestis* (plague) from the American Type Culture Collection. The company that sent him the vials became suspicious after a telephone conversation revealed that his research was purportedly to counteract "Iraqi rats carrying supergerms." One week later, federal and local law enforcement officials obtained a search warrant and found three vials of pathogens in his car and explosives in his home. At the time, there were no laws prohibiting the possession of such agents, and Harris was sentenced to 18 months of probation for wire fraud.[28] This event contributed to the passage of the Antiterrorism and Effective Death Penalty Act of 1996 (Select Agent Legislation), which tightened regulatory controls on the transfer, packaging, and acquisition of pathogens, requiring that all laboratories transferring select agents be registered with the CDC.[29]

Common fermentation techniques used for producing antibiotics, vaccines, foods, and beverages can be used to grow large quantities of biological agents. Simple materials available at home are also suitable for generating the quantities of organisms needed.

There are also some disadvantages to using biological agents as weapons. For instance, hazards are posed to the user during their production and deployment. In addition, they can be inactivated by solar irradiation (and under other climatic conditions) or by chlorine and the treatment process.

There are a wide variety of bacterial agents that could be deployed. Appendix A contains a list published by the CDC, detailing the organisms that they consider to the highest risk for use in a terror attack.

Of the organisms on the CDC list, one in particular that receives a tremendous amount of attention is anthrax. This is due to the mail-borne attacks of 2001, when letters containing anthrax spores were sent to several media outlets and the offices of the U.S. Senate. This incident resulted in five casualties and caused much worry throughout the country.

The disease anthrax is caused by the spore-forming bacteria species *Bacillus anthracis* (fig. 4–18). The spores of these bacteria are extremely hardy and can survive in the environment for years. The spores are resistant to chlorine and thus would be a good candidate for a water agent. There are, however, some problems with its use as an agent in water. The infectious dose by the oral route is thought to be significantly higher than the 8,000–50,000 spores needed to infect by inhalation. Anthrax is easy to grow in culture, but it is much more difficult to grow and purify the spores. The nonspore vegetative cells are sensitive to chlorine even though the spores are resistant. It is unlikely that anyone who has gone to the trouble of purifying anthrax spores in volume would choose to deliver them via water. They are not extremely effective as oral pathogens, and much better use could be made of the spores by deploying them through inhalation or skin contact.

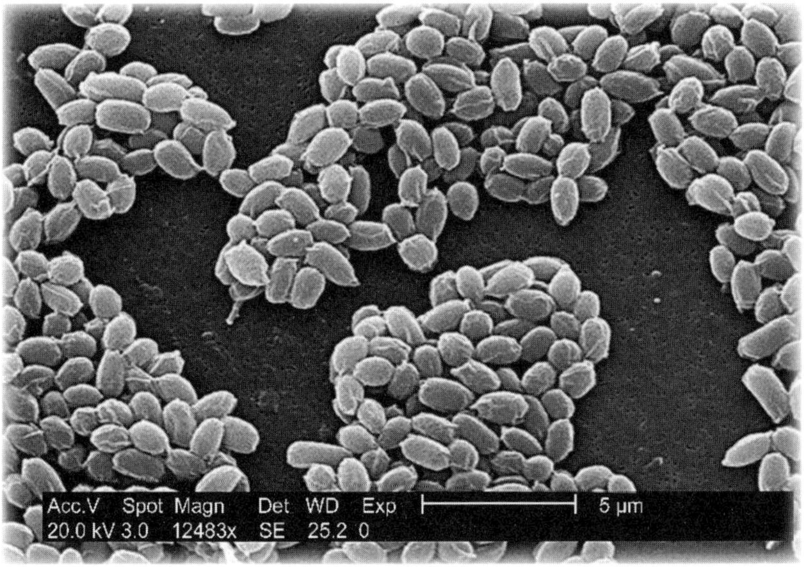

Fig. 4–18. An electron micrograph of spores from the Sterne strain of ***Bacillus anthracis****. These spores can live for many years, enabling the bacteria to survive in a dormant state. While greatly feared—with good reason—these may not be a good choice for water delivery. (Photograph courtesy of Laura Rose and the CDC)*

A much more likely scenario for a water attack would be to use an organism such as *Vibrio cholera,* the causative agent of cholera. This organism has an estimated infective dose of about one thousand organisms. It is very easy to grow in a culture that contains as many as 10^8 organisms per milliliter of culture media. This means that it would require only 30–40 mL of culture to contaminate a volume of one million gallons of finished drinking water. These bacteria, however, are very sensitive to chlorine; they would be readily destroyed unless, as with botox, a dose were used that was so large that a substantial amount of the material remained undegraded or unless the chlorine were removed from the water supply with a co-injection of a reducing material such as sodium thiosulfate or ascorbic acid.

There has been some debate over the exact method that a water attack using a biological weapon would take. There has been debate over whether the perpetrators would be more likely to use a purified and washed culture or a bulk culture that still contained the growth media. While either mode of attack is by all means possible, the general consensus is that an attack would take the form of a dirty attack—that is, still containing the growth media.

It is difficult to wash the media from a culture without substantially reducing the number of viable bacteria found in that culture. When cells are washed, they have a tendency to undergo lysis owing to the change in osmotic pressure. Also, the washing and preparation of the cells would require much more handling of the infectious material by the would-be perpetrators and would thereby increase their chances of becoming infected before they could carry out the attack. In general, if such an attack were to occur, the chances are that it would be with a raw, unwashed culture or even raw sewage.

Protozoa and parasites. Many forms of protozoa are known to be a threat to water. Most of us have experienced that feeling late at night when we know that the next few hours will be spent making frequent and profuse trips to use the facilities. Usually we chalk it up to something that we ate, but this may not always be the root cause of our discomfort. More often than not, the glass of water that we had with our meal may be the culprit. What is the mysterious cause of our discomfort? The answer in many cases is *Cryptosporidium,* a small singe-celled protozoan parasite whose name means "mystery spore."[30]

Cryptosporidium is a family of single-celled animals or protozoa, whose presence is, for the most part, ubiquitous in natural waters.[31] It can be found in the intestinal tract of warm- and cold-blooded animals; from there, it finds its way into water supplies via fecal matter. *Cryptosporidium* has been detected in birds, reptiles, fish, cattle, sheep, pigs, goats, cats, dogs, deer, raccoons, foxes, coyotes, beavers, muskrats, rabbits, squirrels, and humans.[32] There are at least four distinct species of the organism,[31] but the species that is recognized as causing illness in humans is *Cryptosporidium parvum.*[33] *Cryptosporidium* was first isolated and described in 1907 by E. E. Tyzzer.[34] It was not until 1976 that *Cryptosporidium* was recognized as a human pathogen,[35,36] and it wasn't until 1985 that *Cryptosporidium* was recognized as a potential waterborne pathogen.[37]

The unique life cycle of *Cryptosporidium* contributes to its role as a widespread and hard to control pathogen. In an aquatic environment, *Cryptosporidium* exists in the form of a tough and very resistant infective capsule known as an oocyst. These oocysts are very small, 4 to 6 μm, and act as a protective carrying case for the infectious material inside. When one of these oocysts is ingested, it passes through the digestive tract until it reaches the small intestine, where through a process known as excystation, it splits open and releases its contents, small wormlike creatures known as sporozoites. The sporozoites invade the lining of the gut, the epithelium, attaching themselves to the epithelial cells. This disrupts the normal functioning of the intestine and results in severe diarrhea, of up to 10 L per day.[38] As the sporozoites mature, they undergo a sexual process by which they form more oocysts, which can further infect the host or which can be passed out in the feces. The oocytes that are contained in the feces can infect other hosts by direct contact, through the food chain, or by entering the water supply. This results in a cycling of the infectious agent through the environment and from host to host.[33]

The gastrointestinal infection caused by *Cryptosporidium* is known as cryptosporidiosis. After the oocysts are ingested, there is an asymptomatic period of 2–12 days. After this period, symptoms begin to occur, including severe diarrhea, stomach cramps, vomiting, and low fever. There is as yet no known cure for the infection.[38] In most people, the disease tends to be self-limiting and usually lasts 10–30 days. However, in people with compromised immune systems—such as patients with acquired immunodeficiency syndrome (AIDS), people undergoing cancer therapy, the very young, and the very old—the disease may continue indefinitely and in severe cases may lead to death.[39,40] There is some debate as to the exact number of oocysts that need to be ingested to result in disease, but for immunocompromised individuals that dose may be as low as one oocyst.[14]

It is the oocyst stage of its life cycle that makes *Cryptosporidium* a particularly troublesome water contaminant. These tough little packages of infectious material are extremely resistant to the environment and can survive for up to 18 months outside a host.[38] Because the oocysts can survive for long periods and travel large distances, transmission is very likely from sources such as livestock manure found upstream from a water source. Studies conducted by the American Water Service Company and other groups have detected *Cryptosporidium* oocysts in 65–97% of raw water samples tested. The tough outer shell of the oocyst also makes the organisms invulnerable to most conventional methods of water purification. Chlorine and other common disinfectants seem to have little effect on the capsules. The small size of the oocysts also makes likely their passage through all but the most effective filtration systems. Oocysts have been detected in up to 17% of the treated drinking water samples that have been tested.[31,42] These special features of *Cryptosporidium* have inevitably led to outbreaks spread from the public water supply.

In April 1993, thousands of Milwaukee, Wisconsin, residents and visitors were stricken with a flu-like illness caused by the miniscule yet potent protozoan *Cryptosporidium*. The outbreak sickened 403,000 people and was linked to inadequate treatment of drinking water from Lake Michigan. The outbreak resulted in over 400 hospitalizations and 100 deaths. No specific source of the *Cryptosporidium* was ever identified, but runoff from abnormally heavy spring rains most likely carried the protozoa to the lake from a variety of sources.

It would seem that these organisms would make ideal candidates as terror agents. However, one of the drawbacks would be culturing large quantities of the material in a pure form. It would be much easier to simply use extracts of infected animal feces; these, however, would be detectable in many cases. Also, these organisms tend not to kill average healthy adults. Usually a person in moderately good condition can withstand an infection with nothing worse than a case of diarrhea. Fatalities usually occur only in malnourished and immunocompromised individuals. While these organisms are a definite threat, they are probably not the ideal candidates from a terrorist standpoint.

While common waterborne protozoa may be of interest to terrorists, other more exotic protozoa and parasites may be more important agents of concern. A recent article in the *American Society of Microbiology News,* entitled "Raccoons, Parasites Have Bioterrorism Potential," details a parasite known as *Bayliascaris procyonis.* This parasite is a nematode that is found in the intestinal tract of North American raccoons. Human ingestion often leads to infections of the brain, potentially causing brain damage and death. An infected raccoon's feces contains up to 250,000 eggs per gram, and the infectious dose may be as few as 5,000 eggs (see fig. 4–19). The eggs are very resistant to disinfection from chlorine and other methods and can stay viable for years.[43] Raccoons have a tendency to defecate consistently in the same general area, leading to the accumulation of large piles of droppings. Therefore, the raw material for an attack using this parasite is readily obtainable in large quantities at little or no cost. Although off the general radar, *B. procyonis* and similar organisms are definitely viable threats that could be used in water attacks.

Fig. 4–19. Could raccoon feces be a viable terrorist threat agent? Some experts think so.

Viruses. There are a number of viruses on the CDC list. The most feared among these is variola virus, the causative agent of smallpox.

Smallpox has been eradicated from nature, but stockpiles of the virus persist in various locations. There is great fear that these stockpiles could fall into the hands of terrorists. Routine vaccination against the disease was stopped in 1980; thus, much of the U.S. population would be susceptible. There is little known about the virus's infective dose via the oral route, and even less is known about its susceptibility to chlorine or its stability in water. It is doubtful that a terror group that obtained this agent would attempt to use it via an untried and possibly ineffective route. Therefore, more likely viral candidates are viruses such as Rotavirus or the Norwalk virus that are known to be waterborne pathogens. These viruses however are rarely fatal except to the very young and would not make ideal candidates for terrorist agents.

Other extremely lethal viruses, such as Marburg and Ebola, have been deemed to be possible threat agents. Viruses are much more difficult to culture than bacteria and require a degree of sophistication not liable to be present in very many terror organizations. They require cell culture to propagate in volume, and people working with them can be exposed quite easily. Even highly trained individuals with access to biocontainment facilities are not always successful in safely and efficiently culturing these organisms. The terrorists would require cultures prepared by some other, most likely state-sponsored, organization, or they would have to make use of sophisticated personnel and facilities to prepare the agent. These factors make a viral attack on water unlikely.

Danger of a backflow. Overall, the threat of a backflow attack is a very grave danger. Terrorists such as those involved in the Rome incident, described in chapter 3, are obviously aware of this mode of attack and are beginning to orchestrate operations around it. I can think of no other method that is more cost-effective, technologically easier to mount or would result in more casualties and economic damage than a backflow attack on a water system. Some have suggested that an aerosol attack on the heating, ventilation, and air-conditioning (HVAC) system of a building is equally likely. This type of attack would be limited in scope, and as the U.S. Air Force/Colorado State University study shows, the belief that a water attack must be localized is now known to be a fallacy.

A backflow attack can be a very cost-effective method of exacting mass casualties within a population. Imagine a scenario in which a city of 10,000 people is subjected to a backflow attack in winter, with average daily water use of about 100 gallons per capita (thus, calculations are made for 1,000,000 gallons); assume that the attack affects the entire city, at a level inducing 50% fatalities (or 5,000 people), with many more experiencing less severe effects; that each person drinks one liter of water per day; that the majority of the cost of an attack is from the agent; and that pump and plumbing are negligible (less than $200). Table 4–8 indicates volume and cost for some threat agents per death. The specific agent names have been removed and replaced with class names for security reasons.

Table 4–8. The cost-effectiveness of backflow attacks

Product	Cost	LD-50 per 70 kg man 1 liter	Max Cost per Gram	Max Cost per LD-50 Dose	Max. Cost for 1x10^6 Gallons of Water (City of about 10,000)	Cost per Casualty Assuming 50 percent Lethality	Volume of Material
Nematocide	$65/gallon (easily stolen from Ag chemical distributors)	350 mg	$0.07	$0.025	$94,039.35	$18.81 to $0.04 Purchased vs. stolen material	26 (55 gallon drums)
Plant extract	$1 per 15,000 seeds (bought on eBay could collect or grow own for less)	100 seeds	NA	$0.007	$25,000.00	$5.00 to 0.04 purchased vs. grown or gathered	2500 kg
Plant extract	$1 per 500 seeds (bought on eBay could collect or grow own for less)	1 seed	NA	$0.002	$7,500.000	$1.50 to 0.04 purchased vs. grown or gathered	375 kg This is for crude seed prep. Purified material would be about 0.8 kg
Insecticide	$6.70 per kg bulk from China. (40% solution easily stolen from Ag chemical distributors or old farm sources)	62 mg	$0.01675	$0.001	$3,750.00	$0.75 to 0.04 purchased vs. stolen	281.25 kg
Industrial chemical	$13 per kg (easily stolen from industrial sites)	300 mg	$0.013	$0.0039	$14,265.00	$2.85 to 0.04 purchased vs. stolen	1125 kg
Pesticide	No longer commercially available in U.S., but can be found in many agricultural areas and other countries. Easily stolen.	140 mg	NA	NA	NA	0.04 if material is stolen	525 kg
Street drug	Common street drug easily made.	0.0001 mg	$75.00	For effect, not death $0.0075	$28,125.00	100% effected $2.81	375 grams
Plant extract	Material easy to obtain or purchase.	2.1 mg of pure substance or about 7 seeds	$0.57	0.0012 extracted from bulk purchased seeds	$4,410.00	$0.882 to 0.04 purchased vs. if material grown or gathered	7.9 kg

Cyber Attack

Every year, society becomes more dependent on computer technology to facilitate everyday activities. The water industry is by no means an exception to this trend. Water utilities depend on computer technology in everything from billing to the dosing of treatment chemicals. The degree of computer reliance varies greatly by utility; the general trend is for larger systems to be more reliant on technology. This is, however, not a hard and fast rule, and there are many exceptions. As the emphasis on security permeates the industry, it is be expected that computer technology will play an ever-increasing role, because many of the security enhancements being implemented are based on computer technology. This includes everything from advanced SCADA systems controlling sensors capable of detecting contamination, to reverse 911 systems used to notify customers of potential problems. As this reliance on computers increases, so will the system's vulnerability to attack via these routes.

Undergoing a barrage of cyber attacks has become a way of life for most people. Almost everyone has experienced the annoyance of viruses and worms infecting their computers. Terrorists are definitely interested in the Internet and hacking. Some people have referred to the War on Terror as the first war featuring battles in cyberspace. The FBI says that the cyber terrorism threat to the United States is rapidly expanding. "Terrorist groups have shown a clear interest in developing basic hacking tools, and the FBI predicts that terrorist groups will either develop or hire hackers," Keith Lourdeau, an FBI deputy assistant director, told the U.S. Senate in 2004. According to John Arquilla, a professor at the Naval Postgraduate School, documents captured in Afghanistan by U.S. forces in 2001 showed that al Qaeda was trying to develop cyber-terrorist expertise.[44]

Computer systems that control the water supply and wastewater systems "have been the targets of probing by al Qaeda terrorists," says Representative Adam Putnam, who cites U.S. law enforcement and intelligence agencies. Over the past 10 years, unwanted intrusions have occurred in some 50 incidents involving automated systems that control important physical equipment through the Internet, says Joseph Weiss, a security consultant in San Jose, California, who laments that "Not enough people are taking this seriously." In 2004 at a hacking conference in Birmingham, England, a hacking expert presented a detailed paper on cracking into water systems.[44]

According to the National Security Agency (NSA), foreign governments already have or are developing computer attack capabilities, and potential adversaries are developing a body of knowledge about U.S. systems and methods to attack these systems. The National Infrastructure Protection Center (NIPC) reported, in January 2002, that a computer belonging to an individual who had indirect links to Osama bin Laden contained programs that indicated an interest in the structural engineering of dams and other water-retaining structures. The NIPC report also stated that U.S. law enforcement and intelligence agencies had received indications that al Qaeda members had sought information about control systems from multiple Web sites, specifically on water supply and wastewater management practices in the United States and abroad.[45]

Experts in the water industry consider control systems to be among the primary vulnerabilities of drinking water systems. A technologist from the water distribution sector has demonstrated how an intruder could hack into the communication channel between the control center of a water distribution pump station and its remote units, located at water storage and pumping facilities, to either block messages or send false commands to the remote units. Moreover, experts are concerned that terrorists could, for example, trigger a cyber attack releasing harmful amounts of water treatment chemicals, such as chlorine, into the public's drinking water.[45]

Richard Clarke, the former White House head of counterterrorism, says "There is potential vulnerability throughout industry where control systems are connected to the Internet." This problem is not getting much financial attention. The new Transportation Security Administration was awarded $8.5 billion of contracts in 2004, while the DHS's cybersecurity division, by contrast, will have a 2005 budget of less than $80 million.[44]

Regardless of whether they are motivated to commit acts of terrorism, hackers are obviously able to disrupt the system. The severity of disruption depends directly on what control functions a given utility has automated using the Internet. A system that uses the Internet only for billing is at far less risk than one that relies on it for the dosing of fluoride. The threat of cyber attack in most cases would result in little more than annoyance or a denial of service, but there are rare instances in which actual loss of life could result from such an attack. Also, if security functions are routed over the Internet, it is not impossible that a cyber attack could be linked to a backflow attack in a mass-casualty scenario. The threat of cyber attack is real and needs to be addressed.

Subsidiary Infrastructure

The supply of safe drinking water is not a stand-alone enterprise. It relies on other infrastructure components to operate safely and effectively deliver drinking water to the public. Most water utilities rely to some extent on electricity to power pumps and operate treatment systems. As has been shown by the insurgency in Iraq, the power grid is a recognized target easily assaulted by terrorists.

Power supplies are relatively easy to disrupt through physical attack or cyber attack. A disruption in the electric supply grid can rapidly shut down water treatment processes. Even a brief disruption in the power supply can allow tainted water to enter a system. In July 2005, more than 60,000 residents of northwestern Indiana were told to boil their water after a brief power outage affected the company's plants and caused water service problems in Gary, Chesterton, and Portage. According to Roger Swafford, "Customers may experience some dirty, discolored water and were ordered to boil water as a precaution." This sort of disruption occurs regularly. When large power outages occurred in the northeastern United States in 2003, many large cities were left without water, including Cleveland, Ohio, where large electrical pumps were used to move water uphill from Lake Erie. Cleveland was without water for some time, and the incident was referred to as "the worst water crisis in the city's history."[46,47]

Transportation is also crucial to the operation of treatment plants. The water industry relies on rail and truck transport to deliver the chemicals and materials needed in the water treatment process. Any disruption in these delivery systems could lead to the delivery of substandard water to the consumer.

The reliance of the water treatment process on chemicals leads to another hazard. These chemicals themselves are in many cases hazardous—for example, concentrated acids, concentrated bases, fluoride, and chlorine. Most systems in the United States rely on chlorination to disinfect the drinking water supply during final treatment. Wastewater plants also disinfect final effluent with chlorine. Many plants rely on chlorine gas as the source owing to its low cost.

In many cases, the volume of chlorine used is huge, and the chemical is delivered to the treatment pants by rail in a tank car. This is, in fact, a great hazard. Chlorine gas is very poisonous and was used as a warfare agent during World War I. Theoretically, if properly administered, a single tank car of chlorine gas contains enough lethal doses (more than 100 billion) to kill every man, woman, and child on Earth (fig. 4–20). Thankfully the solution to pollution really is dilution in this case. Any ruptured tank's contents are quickly diluted by and attenuated in the atmosphere. The potential of a disaster involving one of these tanks is, however, quite serious. A number of major cities, including Washington, D.C., have attempted to ban the transport of these and other hazardous materials through their cities by rail. The outcome of such attempts is still pending in the courts.

Fig. 4–20. In theory, a single chlorine tank has enough chlorine to kill everyone on Earth. (Photograph courtesy of Shelburne Falls Trolley Museum)

Food and Bottled Water

When apprised of the threat to our water supply, many people shrug it off with the comment, "I drink bottled water." The problem with this is that many of the potential agents that could be used to contaminate a system are not only oral toxins but can also be absorbed through the skin or inhaled. Therefore, taking a shower or turning on lawn sprinklers may be enough to expose someone to a lethal dose. Also, many bottled water sources are simply municipal sources and may or may not undergo further treatment before being placed in a fancy package. Some brands of bottled water are merely bottled tap water.

Even if the water in the bottle is not contaminated, the long supply chain before the product reaches the customer may be vulnerable to terrorist assault. The long history of consumer-product tampering should not be ignored. The Tylenol poisoning incidents, in which bottles of the painkiller were tampered with and resulted in several deaths, come to mind. Therefore, a reliance on bottled water by no means ensures immunity from an assault on the water supply.

Food preparation in the home also often involves the use of water and can thus lead to the ingestion of toxins. Almost all food products are processed with or contain water, often from a municipal source. Any contamination or toxicants introduced into the water supply that feeds such operations could readily find its way into final food products, creating another avenue of exposure.

Conclusion: The Water Supply Is Vulnerable

This chapter has attempted to detail the many and various areas where our water supplies are vulnerable. The superficial analysis that was done immediately after 9/11 indicated that this was not the case and that our water supplies were secure. However, as a more in-depth and detailed analysis was performed, glaring vulnerabilities emerged. The system is vulnerable to a variety of types of attack. Some of these fit the criteria of international terrorism, whereas others are more along the lines of actions that would be perpetrated by a disgruntled employee.

Of all of the vulnerabilities, the one that is the most frightening is the vulnerability to a contaminant attack on the distribution system via a backflow event. This is the type of attack that is the easiest to perpetrate and has the greatest likelihood of resulting in a mass-casualty event. The system can be breached with little effort and little cost. Thus, this type of attack appeals to radicals, nut cases, and people with a grudge. The threat agent used may vary, but the mode of attack would be the same (a backflow attack).

As a mental exercise to determine where the priority terrorist threats are, I performed a Kepner-Tregoe (KT) risk analysis of various threat scenarios for water, taking into account such things as cost of attack, skill levels needed, and potential to cause mass casualties. The rankings in table 4–9, though subjective, are based on my knowledge and experience. The vast majority of the money spent on water security to date has been to prevent attacks on areas and via means that are not as great a risk as backflow attack on the distribution system.

Table 4–9. KT risk analysis of various attack/threat scenarios for water

Site of Attack/ Attack Type	Cost	Mass Casualties >1000	Skill Level Required	Cleanup and Repair Costs	Difficulty gaining access and mounting attack	Total Risk Score > Score = > Risk
Source Water/ Physical	2	2	1	2	2	8
Untreated Water Transport/Physical	8	1	7	4	8	1792
Distribution System/Physical	9	1	10	2	10	1800
Finished Water Storage/Physical	9	1	9	6	9	4374
Untreated Water Storage/Physical	5	4	6	7	6	5040
Source Water/ Contamination	4	6	8	6	6	6912
Treatment Plant/ Physical	8	4	6	7	7	9408
Untreated Water Storage/ Contaminant	7	6	8	5	8	13440
Untreated Water Transport/ Contaminant	8	6	8	5	8	15360
Treatment Plant/ Contaminant	9	9	8	7	8	36288
Finished Water Storage/ Contaminant	9	9	10	10	9	72900
Distribution System/ Contaminant	10	10	10	10	9	90000

When the same analysis is performed on various non–water-related threat scenarios, water attack still comes out near the top (table 4–10). Very little of our national expenditure on security has been used to address water. Perhaps an alteration of our priorities is needed. We need to take every effort to secure our water systems. Some actions have been taken to achieve this end; more are planned, and even more are needed. The next chapter will detail security enhancements that have been implemented, are in process, or are still in the planning stages.

Table 4–10. KT risk analysis of various attack/threat scenarios

Site of Attack/ Attack Type	Cost	Mass Casualties > 1000	Skill Level Required	Cleanup and Repair Costs	Difficulty of gaining access and mounting a successful attack	Total Risk Score > Score = > Risk
Thermonuclear/ Attack	3	10	2	10	7	4200
Aerosol/Bioattack	7	5	5	4	6	4200
Conventional Bomb/Attack	10	2	9	3	10	5400
Dirty Bomb/Attack	6	7	7	7	7	14406
Aerosol Chemical/ Attack	8	7	8	5	8	17920
Industrial Chemical/ Attack	9	6	7	4	8	48384
Distribution System/ Contaminant	10	10	10	10	9	90000

Notes

1. Water Infrastructure Network News. http://www.win-water.org/win_news/112702article.html

2. Hickman, Donald C. 1999. A chemical and biological warfare threat: USAF water systems at risk. Counter Proliferation Paper No. 3. USAF Counter Proliferation Center, Air War College.

3. Burton, John. 2002. Malaysia puts the screws on Singapore over water. *The Financial Times.* March 6.

4. Dick, Ronald L. 2001. Statement for the record of Ronald L. Dick before the House Committee on Transportation and Infrastructure Subcommittee on Water Resources and Environment. October 10.

5. Whitman, Christie. 2001. Whitman allays fears for water security: Possibility of successful contamination is small. EPA press release. October 18.

6. Wikipedia. Operation Chastise. http://en.wikipedia.org/wiki/Operation_Chastise

7. Associated Press. 2004. Bomb study convinces exec to close road atop NY dam. *New York Daily News*. September 2.

8. Fluoridation Chemical Accidents. http://www.actionpa.org/fluoride/chemicals/accidents-us.html

9. EPA. Finished water storage facilities. http://www.epa.gov/OGWDW/tcr/pdf/storage.pdf#search='finished%20water%20storage'

10. U.S. House of Representatives. 2004. Controlling bioterror: Assessing our nation's drinking water security. Hearing before the Subcommittee on Environment and Hazardous Material of the Committee on Energy and Commerce, 108th Congress. September 30.

11. GAO. 2003. Drinking water security: Experts' views on how future federal funding can best be spent to improve security. Report no. GAO-04-29. October.

12. EPA. 2002. Potential contamination due to cross-connections and backflow and the associated health risks: an issue paper. www.eap.gov/ogwdw/tcr/pdf/ccrwhite.pdf

13. Allman, Timothy, and Kenneth Carlson. 2005. Modeling intentional distribution system contamination and detection. *Journal of the American Water Works Association.* 97 (1): 58–61. Although the executive summary of this article is still available on the AWWA Web site, the full text has been removed for security reasons; details of the article could be of potential use to would-be terrorists.

14. Kroll, Dan. 2004. Utilization of a new toxicity testing system as a drinking water surveillance tool. In *Water Quality in the Distribution System.* Edited by William C. Lauer. Denver: AWWA Press.

15. Milstein, Michael. 2004. Wolf poison raises alarms about its terrorism potential. *The Oregonian.* November 3.

16. Brown, R. 1967. Psychedelic guide to the preparation of the Eucharist. http://leda.lycaeum.org/Documents/LSD_Synthesis.8774.shtml

17. Huang, R. Drugs in the anti-terrorism campaign. Center for Disease Information. Washington, DC. http://www.cdi.org/terrorism/narcotics.cfm

18. Mangan, Dan. 2005. Bin Laden cocaine plot fell through. *New York Post.* July 26.

19. Burrows, W. Dickinson, and Sara E. Renner. Biological warfare agents as threats to potable water. Health promotion and preventative medicine form, U.S. Army Center. Aberdeen, MD.

20. Beverly, Robert. [1705] 1949. *The History and Present State of Virginia.* Chapel Hill, NC: University of North Carolina Press.

21. Wannemacher, R. W., R. E. Dinterman, W. L. Thompson, M. O. Schmidt, and W. D. Burrows. 1993. Treatment for removal of biotoxins from drinking water. Report no. TR9120, AD A275958. Fort Detrick, MD: Army Biomedical Research and Development Laboratory.

22. Finegan, William. 2005. Terrorism threat assessment. Presentation at the 3rd annual TICs and TIMs Symposium. Richmond, VA.

23. Budavari, Susan, ed. 1989. *The Merck Index,* 11th Edition. Rahway, NJ: Merck and Company.

24. National Safety Council. How to prevent poisoning in the home. http://www.nsc.org/library/facts/poisong.htm

25. McCann, John. 2003. Water farce. *The Evening Times*. February 21. http://www.eveningtimes.co.uk/hi/news/5013096.html

26. McGovern, Thomas W., and George W. Christopher. 1995–2000. *Biological Warfare and Its Cutaneous Manifestations: The Electronic Textbook of Dermatology.* http://www.telemedicine.org/BioWar/edit_biologic.htm

27. North Atlantic Treaty Organization. 1996. *NATO Handbook on the Medical Aspects of NBC Defensive Operations. Part II—Biological.* NATO Amed P-6(B).

28. Kellman, B. 2001. Biological terrorism: Legal measures for preventing catastrophe. *Harvard Journal of Law and Public Policy.* 24 (2): 417.

29. National Academy of Sciences. 2004. *Biotechnology Research in an Age of Terrorism: Confronting the "Dual Use" Dilemma.* Washington, DC: National Academies Press. pp. 43–78.

30. Goldsmith, P. 1996. H_2Opportunities. *Process Engineering.* 77 (2): 46–47.

31. Roefer, P. A., J. T. Monscvitz, and D. J. Rexing. 1996. The Las Vegas cryptosporidiosis outbreak. *Journal of the America Water Works Association.* 88 (9): 95–107.

32. Dubey, J. P., C. A. Speer, and R. Fayer. 1990. *Cryptosporidiosis of Man and Animals.* Boca Raton, FL: CRC Press.

33. Current, W. L. 1987. Cryptosporidium: Its biology and potential for environmental transmission. *Critical Reviews in Environmental Controls.* 17: 21–51.

34. Tyzzer, E. E. 1907. A sporozoan found in the peptic glands of the common mouse. *Proceedings of the Society for Experimental Biology and Medicine.* 5: 12–13.

35. Mannion, J. B. 1994. On TV reporting and parasites. *Journal of the America Water Works Association.* 86 (11): 6.

36. Milne, R. 1989. Parasite in farm waste threatens water supplies. *New Scientist.* 29 (July): 22.

37. D'Antonio, R. G., R. E. Winn, J. P. Taylor, T. L. Gustafson, W. L. Current, M. M. Rhodes, G. W. Gary, and R. A. Zajac. 1985. A waterborne outbreak of cryptosporidiosis in normal hosts. *Annals of Internal Medicine.* 130: 886–888.

38. Lewis, S. A. 1995. Trouble on tap. *Sierra.* 80 (4): 54–58.

39. Logsdon, G. S., D. Juranek, L. Mason, J. E. Ongerth, C. R. Sterling, and B. L. P. Ungar. 1988. Roundtable: Cryptosporidium. *Journal of the America Water Works Association.* 80 (2): 14–27.

40. Okun, D. A. 1996. From cholera to cancer to cryptosporidosis. *Journal of Environmental Engineering.* June: 453–458.

41. Gurwitt, R. 1994. Something in the water. *Governing.* September: 32–38.

42. LeChevallier, M. W., and W. D. Norton. 1995. *Giardia* and *Cryptosporidium* in raw and finished water. *Journal of the American Water Works Association.* 87 (9): 54–68.

43. McSweegan, Edward. 2002. Raccoons, parasites have bioterrorism potential. *ASM News.* 68 (11): 539–540.

44. Lenzer, Robert, and Nathan Vardi. 2004. Cyber-nightmare. *Forbes.* September 20.

45. GAO. 2004. Critical infrastructure protection: Challenges and efforts to secure control systems. Report no. GAO-04-354. March 30.

46. Mabin, Connie. 2003. Cleveland still coping with water crisis. Associated Press. August 15.

47. *Chicago Tribune.* 2005. Power outage prompts advisory to boil water. July 20.

5

Physical and Plant Security

I have six locks on my door all in a row. When I go out,
I lock every other one. I figure no matter how long somebody stands there
picking the locks, they are always locking three.

—Elayne Boosler

Physical Security in General

When we feel that we are threatened or endangered, the natural reaction is to lock the doors and bar the windows. The water industry's sense of being threatened after 9/11 was no different, and the immediate reaction was to invest in locks and gates. Initially after 9/11, the focus was on the enhancement of physical security for the water supply systems of the nation. Physical security measures make it more difficult for an adversary to gain access to areas or components of the system that could be attacked. Prior to 9/11, security along the water supply network was quite lax, and frequently, security was nonexistent at points that could conceivably be attacked.

Reservoirs, storage facilities, treatment plants, booster stations, and pump houses were often guarded by nothing more than a simple padlock, if at all. Many of the vulnerability assessments mandated by the Bioterrorism Act of 2002 focused on the obvious lack of physical security at these sites. Much of the initial outlay of funding to address security issues was diverted in this direction, and many improvements were and are being made.

Physical security should be considered the first line of defense for all of the components that make up the water supply infrastructure. Experts in security, along with the American Water Works Association (AWWA), the American Society of Civil Engineers (ASCE), and the Water Environment Foundation (WEF), recommend a tiered approach to physical security. The military nomenclature for such a layered approach is "protection in depth." This approach requires that the attacker penetrate a series of security hurdles before reaching the target area. An analogy would be the layers of an onion that must be peeled off before reaching the center. The different layers of the overall system can be enhanced as needed, to guard against whatever threat seems likely.

The AWWA divides a location's upgradeable security assets into four distinct layers:

1. The perimeter of the facility typically includes the fence and access gates that surround the site. The perimeter is considered the first line of the physical security system that, through operational practices, can be sufficient for basic threats such as poorly equipped vandals and criminals.

2. The site is the area between the perimeter and the buildings, structures, and other individual assets. This area provides a unique opportunity for early identification of an unauthorized intruder on the site and initiation of early response.

3. The buildings and structures within a facility, such as a treatment plant or pump station, provide the next physical barrier for stopping intruders. The discussion of buildings and structures is limited to the external features, such as doors, windows, walls, materials, and skylights.

4. Building systems refer to the internal features of buildings and other structures that can protect critical assets or processes from intruders. Examples of these types of features include internal walls and doors, equipment cages, and redundant equipment.

The extent to which each of these layers can or should be hardened with physical security is determined by the threat level and likely modes of attack at a given facility (fig. 5–1).[1]

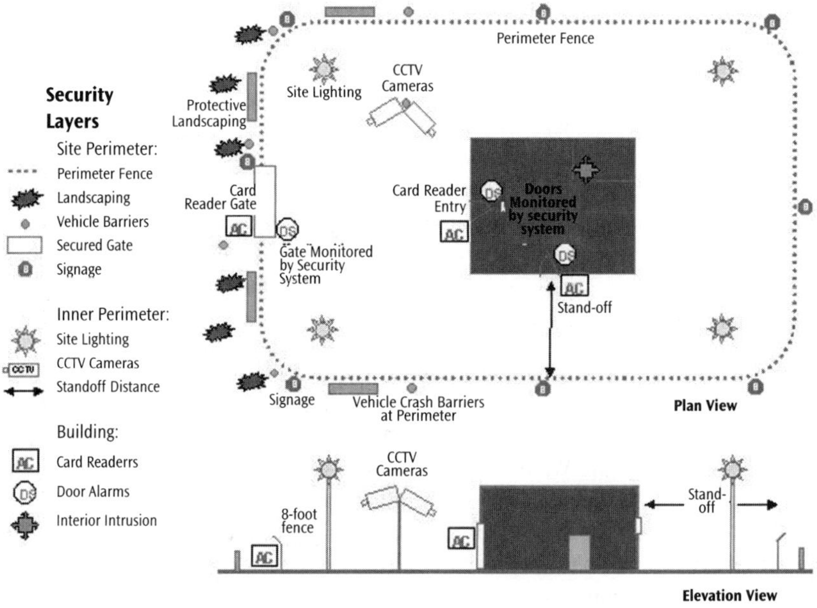

Fig. 5–1. The various layers of physical security

Each component of the water supply network offers opportunities for and challenges to the deployment of physical security. From source water to the end user, there are options for deployment. Each system component will be discussed along with options for enhancing physical security.

Source Water

As discussed in chapter 3, the likelihood of a successful terrorist attack on source waters is remote. Such an attack is, however, not impossible, and due vigilance is warranted to deter or detect an attack. The nature of such source waters as lakes, rivers, and streams makes them very difficult or impossible to physically secure. As previously discussed, the dual-use aspect—for boating, fishing, swimming, transportation, receiving waters for waste streams, and so on—makes limiting access difficult or impossible. Therefore, surveillance of the water quality through monitoring can help in detecting any contamination incident. Monitoring is discussed further in chapter 7.

While physical security is in many cases impractical, some municipalities have isolated their water sources and denied the public access to enhance security and ensure source water quality. This is accomplished through fencing or the use of signage and regular patrols by watchmen or guards. This is a viable option in some but not all cases. Wherever this type of isolation is not possible, it is good practice to maintain regular checks on the sources by water department personnel.

It is wise to rely on help from the public in spotting any unusual activity that may be associated with terrorist activities. Active solicitation of the public's support is always a good idea. Some utilities have personally recruited people who live nearby to keep an eye on key sources. The public should be kept informed of what to look for and how and where to report suspicious activities. The strategic placement of posters (e.g., see fig. 5–2) and the inclusion, in billing statements and other communications, of tips on what to look for and how to report suspicious activity are good ideas. An example of a reporting form is shown in appendix B.

When the public asks what they should be looking for, the blanket statement of "anything out of the ordinary" is often the reply. This is not enough; rather, specific examples of behaviors to look for should be given as well. In general, there are indicators that can be used to recognize potential terrorist groups and potential terrorist probing, surveillance, or pre-attack activities.

Fig. 5–2. A poster from the EPA Drinking Water Security Web site. Posters can be useful tools in telling the public how to report suspicious activity.

Potential terrorists[2]

These four criteria can be used to recognize terrorists, based on patterns of behavior and lifestyle:

- Usually, a group of people (three to seven, or occasionally more) who occupy a house, apartment, or motel rooms and adhere to no obvious or identifiable schedules or pattern of activity indicative of their attendance at work or school.
- A similar group of people who are interested in renting a house, apartment, office space, or storage space without providing a reasonable explanation for its use or appear to be evasive about its intended use. There should be heightened concern if such individuals insist on paying in cash or appear to be carrying large sums of money without credible economic support.

- People who attempt to purchase or lease cars, trucks, or boats with cash, using questionable or insufficient identification or acting evasively about paperwork involved in the transaction.
- Any noticeable chemical smell, especially ammonia or acetone, coming from a neighbor or tenant.

Potential terrorist probing, surveillance, and pre-attack activities[2]

The following are recognizable activities leading up to an attack on source waters or other targets:

- Any theft or losses of badges, credentials, ID cards, or official government, military, or emergency vehicles and uniforms.
- Discovery of anyone possessing false ID in order to access an important, restricted, or sensitive area.
- Photographing, sketching, or performing surveillance of buildings and facilities, especially those involving infrastructure or symbolic structures.
- Trespassing near key facilities or in secure areas, especially by multiple persons and at night
- Uncommon or abandoned vehicles, packages, or containers, especially in crowded or well-traveled areas or in areas next to key infrastructure.
- Uncommon or abandoned vehicles, packages, or containers in the neighborhoods, buildings, or private residences of very important persons, even if no danger is found.
- Observing people who appear to be searching trash containers or placing unusual items in them, such as suitcases, backpacks, or anything out of the ordinary.
- Any theft of sensitive military, government, or facility property such as computers, manuals, directories, and plans.
- Purchases that are made at government surplus sales of military, police, fire, utility, or paramedic vehicles and equipment, especially if there are indications of an intention to refurbish them to working condition.
- The attempted purchase or theft of large numbers of weapons, from guns to knives and everything in between.
- The attempted purchase of large quantities of ingredients and equipment for the manufacture of explosive devices or toxins, especially a frequent or unusual purchase of fertilizer, cleaning supplies, or pesticides.

- Unusual purchases, rentals, or questions concerning pumps and plumbing supplies.
- A noticeable increase or spike in cyber attacks or computer probes within an industry.
- An increase in the number of false alarms, such as fire alarms or bomb threats that require evacuation of a facility.
- Theft of commercial vehicles, such as rental trucks, air-conditioning trucks and vans, delivery truck or vans, or vehicles that can usually gain access to areas otherwise restricted to public traffic. This also includes the theft of government or private company vehicle passes, uniforms, or company handbooks or directories.
- Unknown or unidentified workers attempting to gain access to facilities for repairs, installation of equipment, telephone work, cable installation, construction, or remodeling.
- E-mails that request information regarding details of your facility, personnel, or standard operating procedures, regardless of their origin. Verify everything because e-mail headers can be spoofed.
- Unusual patterns of seemingly unimportant and unrelated activity—for instance, the same person(s) parking near facilities or power plants and fishing near power or water plants and bridges and, in an office setting, people who are unrelated to the office building loitering in the lobby or parking garage.
- Unknown persons or occupied vehicles in the vicinity of a potential target over an extended period of time.

According to the EPA, examples of suspicious activities specifically related to water incidents that should be reported include[3]

- Dumping or discharging material to water sources.
- Climbing or cutting a utility fence.
- Unidentified truck or car parked or loitering near a waterway or facilities for no apparent reason.
- Suspicious opening or tampering with manhole covers, buildings, or equipment.
- Climbing on top of water tanks.
- Photographing or videotaping utility facilities, structures, or equipment.
- Hanging around locks or gates.
- Vehicles other than fire trucks hooked up to hydrants.

Reporting an incident

People should be advised not to confront the suspicious individuals, but rather to report the incident. Important information to include in a report is:[3]

- Nature of the incident
- Your identity and location
- Location of activity
- Description of any vehicle involved (color, make, model, plates)
- Description of suspicious individuals (number, sex, race, hair color, height, weight, clothing)

Even though relying on civilian input to report incidents may be the only option for some source waters, the average person is not a trained observer and thus is likely to not notice or to be shy about reporting incidents. Studies have been conducted since 9/11 in which unmarked trucks with Middle Eastern individuals were hooking up to fire hydrants to see if anyone would report the unusual activity. No one did!

Untreated Water Storage

Dams and reservoirs are often of the same type and scale as source waters, making physical security difficult. They are also often in the multiuse category, making the denial of public access problematic. However, one difference between them and source waters is that dams and reservoirs are usually smaller, more localized sites that represent the key targets and vulnerabilities. For example, consider the actual dam area of a reservoir. It is a simple matter to deny access to dam surfaces by closing roads across the tops of the dams and restricting access to the waterside areas near the dam by use of floating barriers (fig. 5–3). Also, surveillance apparatuses, such as CCTV, can be used to monitor traffic and intrusion into these areas.

Even when it is feasible to limit physical access, the mechanisms are by no means foolproof and can often be circumvented. Signage can be ignored, fences and locks can be cut, surveillance cameras can be avoided or disabled, and barriers can be breached. While such installations may delay an impending attack, they are unlikely to completely dissuade a would-be terrorist from mounting such an assault. Luckily, as previously discussed, an attack on such a facility would not be the most likely terrorist target in a water supply network. While unlikely, such attacks could occur, and it is always a good idea to perform regular inspection of such areas to look for anomalies; as with source waters, recruitment of the eyes and ears of the public can only help in detecting potential threats.

Fig. 5–3. While it may be difficult to physically secure raw water storage locations, key sites such as dam surfaces can be access restricted in many cases. While access at this dam is restricted through the use of physical barriers, it wouldn't take a lot of effort to circumvent them. (Photograph courtesy of TexasFreeway.com)

Raw water transport and intake

Raw water transport facilities, such as pipelines, aqueducts, and ditches, are often extensive in length and remote in location. This layout makes public reporting of potential problems an unreliable defense strategy. Technology can play an important role in helping to secure these areas. CCTV networks can allow a small staff to monitor an extensive network of water conveyances (fig. 5–4). In addition to water quality monitoring discussed in chapter 7 and regular visits and inspections, this is about the only security that can be applied to these extensive systems.

Key locations along the system, such as pump stations and raw water intakes, can be protected with traditional security (e.g., locks and gates). In designs employing traditional protection, it is possible to spend a vast quantity of money without significantly increasing security unless close attention is paid to possible threat scenarios when determining design criteria. For example, the raw water intake pictured in figure 5–5 is surrounded by a fence and a locked gate. It is located only a few yards from a public highway and is down a slight incline from the road. The fence is probably adequate for the original design purpose of keeping people from accidentally falling into the inlet and drowning, but it would be inadequate in preventing an intentional contamination event in which a tanker truck full of a toxic liquid simply unloaded its cargo down the incline and allowed it to flow into the inlet, thus contaminating the water heading for the treatment plant.

Fig. 5–4. The use of CCTV surveillance allows a single operator to monitor a wide geographic area.

Fig. 5–5. The security at this facility is adequate for past requirements of public safety but would offer little or no deterrence to a contamination attack.

When engineering such security measures, all avenues of attack must be considered, with choices made on a risk-to-cost basis. For example, the enclosure pictured in figure 5–6 protects the intake supplying water from a reservoir to a treatment plant for a large city. The enclosure is state of the art and could probably stop a bulldozer. During its construction, the structure was very expensive and met a lot of public opposition to its unsightly appearance, as the area around the reservoir is a popular scenic walkway. This construction project would probably stop a truck from gaining entrance via the land side but could easily be thwarted by using a hose to deliver material via the water side. Also, reservoirs have a tendency to fluctuate in level over time. A very dry summer could conceivably leave a mode of entry to the water intake that was no longer protected by the fence but was left dry by the receding waters.

Fig. 5–6. This very pricey enclosure protecting a raw water inlet may not be adequate to thwart a smart terrorist or a dry summer, as it ends at the water's edge.

Treatment Plants

Treatment plants are the first—and in some networks the only—component of the system that can make full use of the tiered system of physical security that is advocated by many experts (see fig. 5–1). As static locations with a small geographic area, plants can make use of all four layers of security. It is possible to install a wide variety of equipment and devices to help ensure the security of such facilities. Each layer of security is dependent on the others for its overall effectiveness, but no matter how simple or how elaborate a physical security system is, it might still be breached. Sometimes, it is just as important to know if and when security has been breached as it is to prevent the breach from occurring in the first place.

Perimeter

The perimeter of the water treatment facility can be subjected to access control by the proper use and deployment of physical security features. These include fencing, security-based landscaping, crash barriers, intrusions sensors, and guard posts. Smaller facilities with fewer resources will need to make use of less expensive passive security features, such as traditional gates and locks, while facilities with more capital may opt for higher-tech solutions, such as biometric security locks and intruder detection (for details, see app. C). It is a good idea to prevent access by unauthorized personnel, and it is definitely advantageous to prevent large vehicles from approaching the plant until the operator and cargo have been identified, to prevent total disruption of the facility by a conventional truck bomb.

A wide variety of perimeter fencing and security measures can be employed. From simple cyclone fences to electrified razor wire, the options that best fit a facility are chosen based on cost and risk analysis. Many fencing systems are also equipped with intruder-sensing devices that are activated when the fence is breached or crossed in an unauthorized area. These systems offer a means of notifying the appropriate personnel when a perimeter security system has been compromised. One drawback of these systems is that false alarms may be caused by routine circuit problems or breach through natural causes (e.g., animal intrusion). Newer and more elaborate systems have means to diminish or remove these problems, but with an associated rise in cost.

Grounds

The grounds between the perimeter security and the actual site buildings offer another opportunity for enhancing security. At the very least, this area should be kept clean and tidy, not to mention properly lighted, so that anyone who is not meant to be there can be visually detected. This area also offers the opportunity for deployment of such security measures as CCTV monitoring and motion detectors.

Buildings

Buildings can be secured in a traditional manner with limited access granted through lock and key. Advanced biometric devices that work on fingerprint- or retina-scanning technologies, as well as traditional card scanners, are becoming more inexpensive as their use becomes more widespread. CCTV and traditional hardwired alarms and security systems can play a role in securing buildings.

Interior spaces

Interior spaces provide another area to enhance security. Critical areas and equipment can be secured in access-restricted areas. These can be either locked rooms or cages in larger areas. These areas can be secured with traditional locks or, once again, with key cards; biometrics and security cameras can be used as well.

Finished Water Storage

Finished water storage areas can be made more secure by denying access to unauthorized people through the classic means of fences and locks. Water tanks and towers can also be made safer by physically blocking access to ladders and stairwells used to gain access to the water. This means of denying unauthorized access has often been utilized in the past, to prevent vandalism, and has met with limited success. (See fig. 4–12.) Security cameras, motion detectors, automatically activated lights, and other high-tech advances can augment old-fashioned methods. Moreover, it never hurts to alert local law enforcement officials to the vulnerability of these areas and secure their cooperation in keeping an eye on them.

Furthermore, many of the reservoirs and tanks that are used to store the nation's finished drinking water supplies are uncovered and open to the environment. Facilities such as these are easily accessible to contamination, whether natural or induced by terrorists or vandals. The cost to cover these existing storage tanks can be very expensive. While covering these reservoirs can be financially prohibitive, there are numerous options available as to the types and styles of covers that may be used (fig. 5–7). The enhancement of security and the general improvement of water quality may make these rather expensive improvements worth the expenditure.

Fig. 5–7. There are many options for installing covers on open water storage facilities. (Courtesy of Temcor)

Finished Water Transport—the Distribution System

The distribution system, because of its vast nature, is not very amenable to the deployment of physical security measures. The system is accessible and can be contaminated anywhere along the pipes, which in some cases stretch for thousands of miles. While not every inch of the system can be physically secured, certain key access points, such as pumping stations and fire hydrants, can be made more secure through locking them. Many companies specifically produce secure locking systems designed for hydrants (fig. 5–8).

Fig. 5–8. Two styles of hydrant lock used to prevent unauthorized access to hydrants (Courtesy of McGaurd)

The distribution system has a large number of access points. Many pumping stations, chlorine booster stations, and other equipment are located underground, accessed through manholes. Pipes are also accessible via the routes used for maintenance, such as manholes. Specialized locking mechanisms have also been designed for these structures (fig. 5–9).

Fig. 5–9. There are many options for securing and limiting access to manholes and other access points.

Backflow prevention

The trouble with locking up these key points is that it does nothing to limit the access to the water in the rest of the system. Any point that can be used to obtain water from the system can be used to backflow contaminants into the system. The industry has long been aware of the problem of accidental backflow, and manufacturers have designed a variety of backflow prevention devices to prevent such accidental contamination incidents. These methods and devices are discussed in the following paragraphs.

It is imperative to remember that these devices were designed to prevent accidental contamination. They are physical devices that are located in unsecured locations. As such, they can be disabled or removed by terror-minded individuals. They may offer some respite form an unskilled attack, but offer little protection from a skilled and knowledgeable adversary.

Air gap assembly.[4] An air gap is a nonmechanical backflow prevention method that is effective against backsiphonage or backpressure conditions. An air gap system is implemented by physically separating the supply pipe from the receiving vessel (fig. 5–10). This breaks the pressure between the inlet and the outlet, thereby preventing backflow. According to standard engineering design practice, the distance between the supply pipe and the receiving vessel should be at least twice the diameter of the water supply outlet and never less than one inch. An air gap is acceptable for protection against contaminant or pollutant hazards; in addition, an air gap may be the best means of protecting against accidental contamination with lethal hazards.

An air gap assembly may be purchased as separate components, which are then integrated into existing plumbing and piping configurations (fig. 5–10). Because an air gap breaks the pressure between the inlet and the outlet, a booster pump is needed downstream to ensure pressure, unless the flow of the water by gravity is sufficient for the downstream water use.

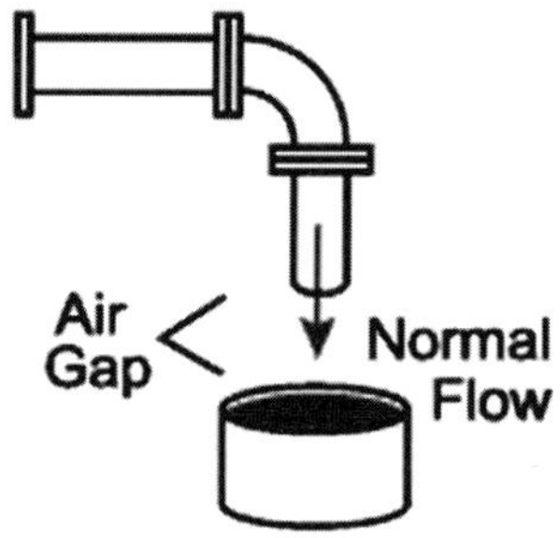

Fig. 5–10. Typical air gap assembly

The air gap drain is a very effective way to prevent accidental contamination of the water system; however, an air gap is not always practical and can easily be bypassed. If the distance between the supply pipe and receiving vessel is compromised either purposely or inadvertently to prevent excessive splash, the air gap is defeated. Also, with an air gap, water is exposed to the surrounding air; therefore, the aspiration effect could potentially drag down airborne pollutants or contaminants into the receiving vessel. Often it is not possible to incorporate an air gap into the design of a system; instead, designers may opt to install mechanical backflow prevention devices, which provide physical barriers to backflow. Physical backflow prevention devices are described next.

Double check valve.[4] A double check valve (fig. 5–11) is a mechanical device that consists of two single check valves coupled within one body and two tightly closing gate valves, one located at each end of the unit. Each check valve consists of a physical plate connected to the top of the pipe by a hinge. The hinge is oriented such that forward flow in the pipe keeps pressure on the plate and keeps it open, permitting the passage of fluid in the intended direction of flow. Thus, under normal conditions, the check valves remain open. In the absence of water flow, the plate is not held open by flow in the correct direction, and the valves close until the normal water flow resumes. In the event of backflow, the flow is against the direction of the hinge, so the plate remains closed.

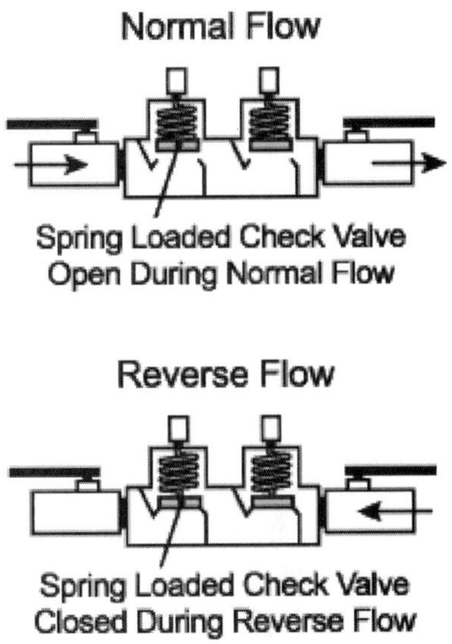

Fig. 5–11. Double check valve assembly

A double check valve may be used under continuous pressure. It can be effective against either backpressure or backsiphonage and may be used to protect against pollutant hazards. Note that double check valves are susceptible to interference from materials within the piping system. For example, grit or fibers can catch under the valves, causing them to remain open and potentially allowing leakage back into the system.

Reduced-pressure (RP) principle assembly.[4] The principle behind an RP principle device is the reduction of a negative pressure differential between the upstream and downstream ends of a line, thereby preventing backflow. An RP principle assembly is a mechanical backflow preventer that is essentially two check valves with an automatically operating pressure relief valve placed in between them. This system is designed such that the zone between the two checks is always at a lower pressure than the supply pressure. Under normal flow conditions, the check valves remain open, and the relief valve is closed.

An RP principle assembly (fig. 5–12) is effective against either backpressure or backsiphonage and may be used to protect against pollutant or contaminant hazards. In the event of backsiphonage, the relief valve will open to allow the induction of air to break the vacuum. In the event of backpressure, the opened relief valve routes the contaminated water out of the system (drainage can be provided for such spillage). RP principle assemblies may be used under constant pressure and are commonly installed on high-hazard installations. The RP principle assembly also contains two shut-off valves, upstream and downstream of the check valves, and a series of test cocks for periodic testing of the valves.

Pressure vacuum breaker (PVB).[4] The principle behind a PVB device (fig. 5–13) is to break the vacuum created during a backsiphonage event, thereby preventing backflow. A PVB consists of a spring-loaded check valve that closes tightly when the pressure in the assembly drops or when zero flow occurs, plus an air relief valve (located on the discharge side of the check valve) that opens to break a siphon when the pressure in the assembly drops. The assembly also includes two shut-off valves and two test cocks for periodic testing of the assembly. The air relief valve ensures that no nonpotable liquid is siphoned back into the potable water system.

PVBs prevent the backflow of contaminated water into a potable drinking main line, but they are not designed for backpressure conditions. PVBs may be used under continuous pressure, but the air inlet valve may become stuck in the closed position after long periods of continuous pressure. A PVB may be used to protect against backsiphonage only or to protect against pollutant and contaminant hazards as well.

It may be, for all practical purposes, impossible to completely protect against a backflow event. If a backflow cannot be prevented, it becomes imperative to detect such an event as it occurs, so that the proper response actions can be initiated. Monitoring for such events is covered in chapter 7.

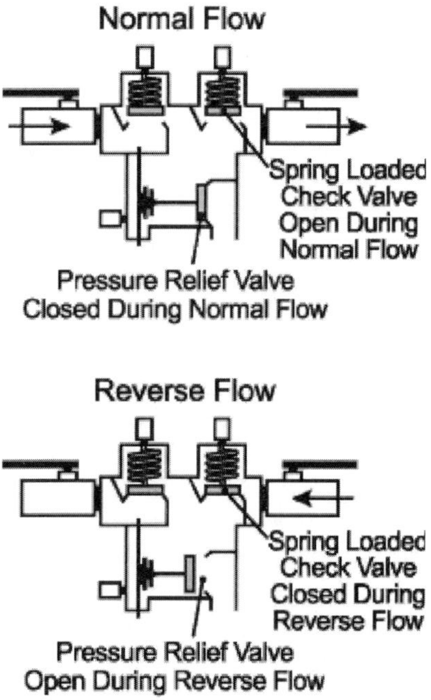

Fig. 5–12. RP principle assembly

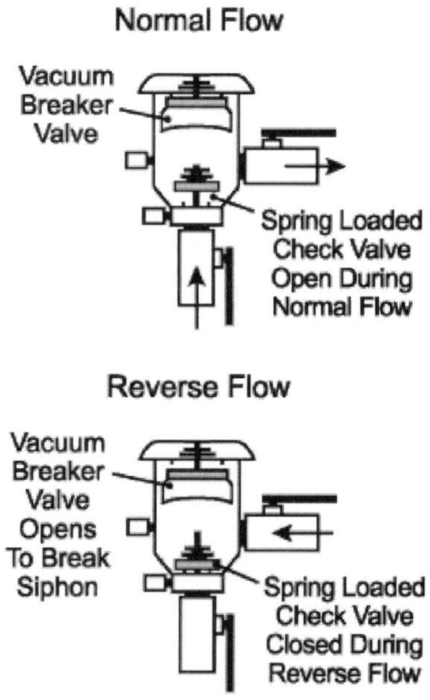

Fig. 5–13. PVB backflow prevention device

Chemicals

The chemicals found in a water treatment plant present a particular hazard from a security standpoint. These materials can be extremely toxic and can represent a hazard by either overdosing in the water or aerosol release. For these reasons, there are some basic security measures that can be taken to guard these stocks.

- Keep the amount of chemicals located on site to a minimum. Agreements should be made with suppliers to deliver the materials as they are needed, so that there are not large stockpiles in or near the treatment plant. This is a common practice in the industrial manufacturing sector and is known as just-in-time (JIT) delivery.
- Agreements should be made with suppliers to provide adequate background checks and security in the delivery chain. During the delivery process, there should be a means of verifying that the delivery person is the person contracted with to do the delivery and not an imposter. As was discussed in chapter 4, one means of attack would be to simply deliver the wrong or adulterated chemicals.
- Simple quality-control checks should be in place, to verify the identity and the quality of the delivered chemicals. Have more than one person check this at different stages to avoid mistakes or deliberate fabrication by an infiltrator.
- Once the chemicals are delivered, they should be stored in a secure, access-limited area. This can be either a locked room or a cage and should be equipped according to the appropriate level of security, including combination locks and biometric recognition devices if appropriate.
- Dosing equipment should also be secondarily contained with access only by the personnel that need to operate it.

Personnel

While it is generally a good idea and a widely accepted practice in the industry not to hire personnel with direct links to international terrorist organizations, this is not where the majority of insider security problems arise. Usually, it is from a disgruntled employee who doesn't fit the mold of a terrorist but may engage in terroristlike actions in retaliation for actual or perceived wrongs in the workplace. The best practice in these cases is to anticipate problems and deal with them in a timely manner; a variety of signs and symptoms can be used to recognize problems before they get out of hand and deal with them through counseling or other means. While these criteria do not give a set profile, just as there is no set profile of a terrorist, they are characteristics that have been exhibited in past incidences of workplace violence.

According to the human resources Web page of Carnegie Mellon University,[5] the general characteristics of a potentially violent worker include

1. Any race or sex, but statistically white male
2. Between ages of 30 and 40 years
3. Low self-esteem
4. Considered a loner, socially isolated
5. Exhibits a disgruntled attitude regarding perceived injustices in the workplace
6. May complain regularly about poor working conditions or an unsatisfactory working environment
7. May complain of heightened stress at work
8. Transient job history
9. Chronic labor-management disputes
10. May cause fear or unrest among coworkers and supervisors
11. May have made threats against coworkers, supervisors, or the organization
12. Fascination with military or paramilitary subjects
13. Gun or weapons collector
14. Excessive interest in media reports of violence, especially in the workplace
15. Unstable family life
16. Demonstrates few, if any, healthy outlets for rage
17. Has requested some type of help in the past
18. Poor temper control
19. Numerous unresolved claims or physical (health-related) or emotional damage suffered on the job
20. History of drug and/or alcohol abuse
21. May exhibit psychiatric symptoms

According to Norman Bates of Liability Consultants, certain characteristics and behaviors become more pronounced and therefore more noticeable before the occurrence of an incident.[6]

1. Unexpected increase in absenteeism
2. Repeated violations of company policies
3. Behavior that seems to indicate paranoia
4. Has a plan "to solve all problems"
5. Depression and withdrawal

6. Noticeable lack of attention to personal hygiene and appearance
7. Explosive outburst of anger or rage without provocation
8. Verbal abuse or threats of coworkers or supervisors
9. Frequent or vague physical complaints
10. Increased unsolicited comments about firearms
11. Resistance and overreaction to policy and procedure changes

While there is no surefire method to anticipate every impending incident, these signs can often warn of the likelihood of an upcoming event. It is best, however, to have set policies with regard to preemployment screening practices, so that those potential employees who exhibit risk factors for violence are not hired in the first place. References, both personal and business, should always be checked; criminal and credit checks may be appropriate for employees in critical functions.

Notes

1. American Water Works Association, American Society of Civil Engineers, and Water Environment Foundation. 2004. Water infrastructure security enhancements: Interim voluntary security guidance for water utilities. http://www.awwa.org/science/wise/#P7_623

2. Hagman, Douglas J. 2005. Identifying potential terrorists and sleeper cells in America. North East Intelligence Network. http://www.homelandsecurityus.com/site/modules/news/

3. EPA. Water security and you. http://www.epa.gov/safewater/watersecurity/pubs/water-security-article.pdf

4. USEPA. 2005. Water and wastewater security product guide. http://www.epa.gov/safewater/watersecurity/guide/backflowpreventiondevices.html

5. Carnegie Mellon University. Recognizing the potentially violent employee. http://hr.web.cmu.edu/leadership/files/WorkplaceViolence.doc

6. Heehan, Maggie Head. 2001. Playing it safe: Experts advise on security awareness. *Texas Technology*. June.

6

CYBERSECURITY

We are at risk. America depends on computers. They control power delivery, communications, aviation, and financial services. They are used to stor[ing] vital information, from medical records to business plans, to criminal records. Although we trust them, they are vulnerable—to the effects of poor design and insufficient quality control, to accident, and perhaps most alarmingly, to deliberate attack. The modern thief can steal more with a computer than with a gun. Tomorrow's terrorist may be able to do more damage with a keyboard than with a bomb.

—National Research Council[1]

Introduction

While I am by no means an expert in cybersecurity, the threat to water from such an attack is real and should be considered here. For an in-depth look at the subject, I recommend *Cybersecurity for SCADA Systems,* by Tim Shaw.

Cyber attack is one form of attack of which we are, unfortunately, too aware. In fact, as I prepare this manuscript, I am having some trouble with formatting owing to a virus picked up by my computer. While this is annoying, it is by no means life threatening, and it is very unlikely that professional terrorism is to blame. Thousands of cyber attacks transpire every day, including viruses, worms, and identity theft; for the most part, these have little effect beyond annoyance or minor data loss.

However, as our reliance on computers increases, to perform functions and operate systems that were not formerly computerized, our vulnerability to attack and the magnitude of the consequences increase. In the water industry, as supervisory and control functions become increasingly more computer based, the chance that these systems will be deliberately attacked or suffer collateral damage from a nondirected attack increases. The vast majority of these attacks are perpetrated by insiders, amateurs, and hackers for fun and games. Very few attacks are perpetrated by professional hackers or are in any way terror related. A breakdown of the characteristics of groups responsible for such attacks is shown in figure 6–1. Even though the vast majority are nondirected attacks, they still must be taken seriously and guarded against, and the menace of a deliberate and directed attack on the water supply network via cyberspace must be considered a valid threat.

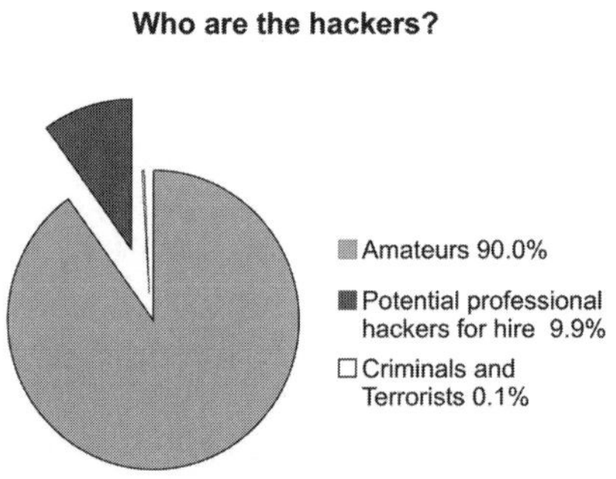

Fig. 6–1. The majority of hacking incidents can be attributed to amateurs and thrill seekers. Very few are professional or terrorist in nature. (After data from IBM Global Security Analysis Lab)

Vulnerabilities

The computer infrastructure utilized by the water utilities can be divided into two separate areas: general information technology (IT) and SCADA. What types of system are in place, how they are set up, and what they are being used for dictate the degree of the vulnerability of a given network. IT systems can be utilized for a variety of functions. Among these are keeping human resource (HR) records, inventory, and delivery schedules; payroll and customer billing functions; maintenance records; laboratory information management systems (LIMS); hydraulic and system models; e-mail; and general data storage of documents, including emergency response plans.

The IT components of a system are often quite vulnerable. These systems are often used for multiple purposes (those listed previously). They are also commonly accessed by a number and variety of employees and have a tendency to be cross-linked—to each other, in a network mode, and to the outside, via the Internet. This presents a number of vulnerabilities. Hackers could gain access to detailed system plans and standard operating procedures (SOPs) via the Internet. This detailed knowledge could be used in formulating plans of attack in an attempt to compromise the system. Personnel records and other information could be used to compromise employees and force them to work with the terrorist. Recall the incident described in chapter 2, in which a federal employee was blackmailed by terrorists into helping with their plans.

In linked systems, an attack on the e-mail system via viruses or worms could disable the operation and affect everything from delivery of needed chemicals to billing operations. While such an attack is not severe in its immediate consequences, it can be costly to correct. Also, the public relations of the utility are bound to suffer if all of the billing statements suddenly show up at the customers' residences with an extra digit or two attached. Laboratory records could also be compromised, resulting in poor water quality and, potentially, in fines from regulators, for not ensuring adequate water quality. Overall, attacks on the IT segment of the utilities' network tend to fall into the nuisance category; however, serious consequences are possible if the system is used to access data used in planning another form of attack.

Attacks on the SCADA portion of the system may be more severe. SCADA systems generally consist of three key components:

1. Field devices or remote terminal units (RTUs)
2. Central or host computer(s)
3. Communications devices or networks

As the term SCADA implies, the host computers allow for supervision and control of the remote site. The majority of the control exercised over the system is executed automatically by the RTUs at the remote sites. The control functions executed by the host normally consist of basic site override or supervisory-level commands. Data acquisition begins at the RTU level and includes meter and sensor readings and equipment statuses that are communicated to the SCADA central host on a set basis. The collected information is then compiled, formatted, and presented such that a control room operator using the SCADA system can decide whether to override normal RTU controls. In more modern systems, even some of these supervisory-type decisions are automated at the host controller level.

As security enhancements to the water supply network are implemented, a number of systems are being configured to feed into existing SCADA systems. This compounds the importance of maintaining security for the SCADA system as a whole. The instruments and meters responsible for maintaining the system are often controlled from the same location as the security apparatus designed to keep watch over these operations. A knowledgable person could conceivably access the SCADA system and disable certain treatment functions while accessing the instruments responsible for detecting such an event and altering their readings to make conditions appear normal.

Because SCADA systems are by definition used to control remote locations from a central access point, communications are integral. A variety of communications options are utilized by SCADA systems. These include Internet connections, hard telephone lines, and radio telemetry. These communications systems are all, to some extent, vulnerable to disruption, failure, or hijacking. Regardless of which mode is utilized, the reliance on communications over a distance makes SCADA systems assailable.

A recent study conducted at several major utilities was able to discern a variety of common vulnerabilities in the system's cyber functions.[2]

- Users of a computer or workstations often fail to log off when no longer using the system. This leaves the computer open for extended periods and accessible to anyone walking by, rendering passwords or other user ID systems ineffectual.
- Physical access to SCADA equipment was not secured, and access could be achieved with ease.
- SCADA could be accessed from remote locations via dial-up modems or the Internet.
- Wireless access points were not secured.
- Most SCADA networks were directly or indirectly connected to the Internet, leaving them vulnerable to hackers.
- Firewalls between the Internet and internal systems were nonexistent, weak, or unverified.
- Records of system events were not monitored.
- Intruder-detection systems to warn of possible hackers, among other intruders, were not used.
- Known problems with software were not routinely patched or otherwise rectified.
- Passwords and configurations were not optimized for security. These were often left on default settings.

There is no doubt that these deficiencies leave the system vulnerable to a variety of cyber attack scenarios. These could result in nuisance-style attacks or the compromising of data used to plan an attack; in the worst-case scenario, in which a SCADA system is hijacked, the operational mechanics of the system could be used to compromise the water quality or prevent the timely and appropriate response to an outside event. There are, however, any number of relatively simple actions that can guard against or mitigate such scenarios.

Securing the Network

The good news about securing the network is that the majority of actions required are procedural in nature and hence inexpensive to implement. The bad news is that securing the system requires changes in behavior and modes of operation by employees. Many of these behaviors are ingrained in the utility's working culture and may be difficult to change, but owing to the vulnerability of the system, it is important to make the effort, stressing to all employees the critical nature of the changes in policy. Detailed in the next section are a number of actions that should be implemented to enhance cybersecurity.[2–4]

General housekeeping

- All systems should be equipped with antivirus software, such as McAfee or Norton.
- Antivirus software should be regularly updated with the latest virus patterns.
- All manufacturer-recommended software patches should be installed as soon after their release as possible.
- Stress to all employees that they should log off the system when not actively using it. Computers should be automatically configured to log off or lock if not in use for a set period of time.
- No unauthorized additions of software or modems should be allowed. This needs to be strictly observed, so that these additions do not open new routes, of which you are unaware, into your system.
- Firewalls should be properly designed and installed between various parts of the system and all outside connections to prevent unauthorized access.
- The SCADA control room, main IT server area, and other critical points in the IT/SCADA system should be strictly access controlled. Biometric devices on locks are appropriate at these critical points.
- Backup systems and/or tape backups of data should be maintained, in the case of system destruction or disaster recovery.
- More than one type of communications system for remote SCADA devices should be incorporated into the system design, to take over in case of the failure of the primary mode of communication.
- Network administrators should be trained to perform and regularly implement audits to detect unauthorized access and use.

Limiting access

- The absolute minimum number of people should be granted access to the system.
- All outside connections should be identified and proper security should be installed.
- Access by vendors or contractors should be identified and monitored.
- Permission for remote access should be limited to those whom really need it.
- All personnel with access, including vendors and contractors, should be screened with background checks.
- Physical access to workstations should be limited if possible. Card keys or biometrics may be used.

Passwords

- All employees with access to the system should have separate, nonobvious passwords. Password should be unique to an individual and should not be shared with a group. It is possible and fairly inexpensive to incorporate biometrics, such as fingerprint identification (fig. 6–2), into the logon procedures. When combined with a password, this makes for a very secure logon procedure. The use of unique logon credentials not only provides security but also allows for a record of who did what when logged on to the system.
- Passwords should never be written down.
- All computers whether at home or at work used to connect to the system should be password protected.
- There should be a set IT chief with explicit policies in place as to who does and does not have access to the system.
- There should be a set policy for granting access and also more importantly revoking access. For example, if an employee is terminated one of the first actions should be to disable that password so that the employee no longer has access to the system. This is standard practice in most industries.

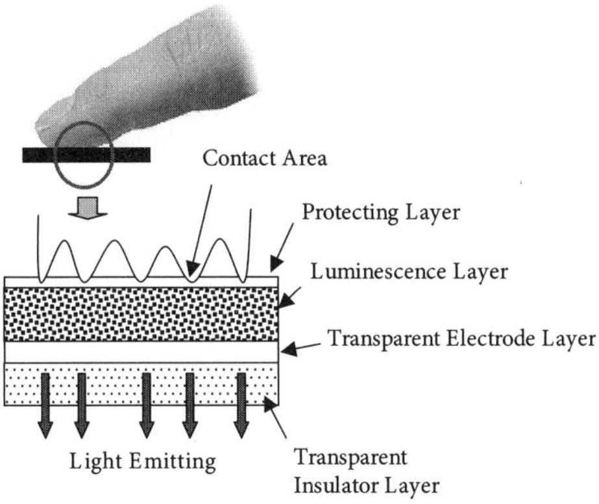

Fig. 6–2. Simple, inexpensive biometric devices such as the fingerprint recognition system pictured here can provide very effective security when used in combination with passwords. (Photograph courtesy of Integrated Biometrics)

Communications

- SCADA systems should not be accessed through the Internet if more secure options, such as wireless radio and hardwired circuits, are available.
- As a rule, telemetry with encryption should always be used, to prevent interception of communications.
- Modems should be configured to allow dial-up access only from a specified set of numbers.
- A timer should be used to turn off modems after a specified time if not in use.
- Wireless access should be limited.

Planning, testing, and audits

- Sign up for and pay attention to a service for reporting new cybersecurity problems (e.g., see the U.S. Computer Emergency Readiness Team Web site [www.us-cert.gov/federal/]).
- All security measures (e.g., firewalls) should be tested by challenging them on a regular basis.
- Network scans for intrusion, password checks, and system maintenance should be performed on a regular basis.
- Regular audits of employee compliance with and understanding of IT security policies should be performed.
- Emergency response and system backup programs should be in place and regularly tested for efficiency of switchover.

Intruder detection

No matter how secure we think we have made the system, there is bound to be someone out there who, with malicious intent or just for the fun of it, will find a way to infiltrate the security system. This is why one of the most important aspects of a security system is intruder detection, allowing you to know when a system has been breached and what areas have been accessed.

Conclusion

A properly organized and executed IT/SCADA security program is integral to maintaining the security of a water operation. Planning and execution is the key to success in this area. Most of the security upgrades that have been discussed are relatively inexpensive and simply require a change in operating procedures. While the likelihood of a true terrorist attack on the cyber part of the water supply network is not high, most systems will at one time or another experience insider or hacker-type nuisance attacks. These attacks, while annoying, tend not to cause any lasting damage. Taking steps to prevent them can result in a more secure system that will not be prone to nuisance attacks. Thus, preparing the system for even a nuisance attack will make the organization ready to thwart a more serious and concentrated attack.

Notes

1. National Research Council. 1991. Computers at risk: Safe computing in the information age. Report from the National Research Council. Washington, DC: National Academy Press.

2. Panguluri, Sirinivas, William Phillips, and Robert Clark. 2004. Cyber threats and IT/SCADA system vulnerability. In *Water Supply System Security*. Edited by Larry W. Mays. New York: McGraw-Hill.

3. Cybersecurity for the Homeland. 2004. Report of the activities and findings by the chairman and ranking member Subcommittee on Cybersecurity, Science, and Research and Development of the U.S. House of Representatives Select Committee on Homeland Security. December.

4. American Water Works Association, American Society of Civil Engineers, and Water Environment Foundation. 2004. Water infrastructure security enhancements: Interim voluntary security guidance for water utilities. http://www.awwa.org/science/wise/#P7_623

7

MONITORING

You can observe a lot just by watching.

—Yogi Berra

Introduction

Would you volunteer to drive a car through heavy traffic while blindfolded? What if you were allowed to peek every couple of hours? What if you could peek every couple of minutes? I don't think many people outside a casting call for the TV show *Fear Factor* could be persuaded to take on such a task. Such a driver would be out of control: according to the old engineering adage, "You must be able to observe a system before you can control it." Yet, we are in effect blindfolded to the minute-by-minute operation of our current water supply systems.

Little monitoring of our water occurs outside treatment plants. Occasional grab samples are taken from water sources and within the distribution system, but this is no better than being allowed to peek from under the blindfold a couple of times per day. This may be adequate in a static system, but it is ineffectual and counterproductive in a dynamic system such as the water supply network. A system being observed in this manner is effectively out of operational control, and—if the goal is to detect unexpected transient events such as contamination actions, whether terror related or accidental—grab samples provide only a false sense of security.

While many of the vulnerabilities of the drinking water infrastructure can be lessened by ramping up physical security and improving policy, some extremely serious vulnerabilities remain unaddressed. The event with the most potential consequences is intentional contamination. Operationally, there is no effective means to completely prevent such an attack. Therefore, the only option is to attempt to detect an attack as soon as possible, to enable the initiation of actions that mitigate the effect. In other words, the only option is monitoring.

The two locations that are practically indefensible from a physical security standpoint and could benefit from monitoring are source waters and the distribution system. Both offer a distinct set of challenges and opportunities as far as monitoring is concerned.

One problem in designing a monitoring system for water is that the vast number of agents that a terrorist could utilize to compromise a water supply system precludes monitoring on an individual chemical basis. Chemical warfare agents (e.g., VX, Sarin, and Soman); commercially available herbicides, pesticides, and rodenticides; street drugs (e.g., LSD and heroin); heavy metals; radionuclides; cyanide; and a host of other industrial chemicals could be exploited as weapons. There are also a variety of biological agents and biotoxins that could be problematic. The possible number of chemical and biological substances that could be used in an attack is very large.

To be truly effective, a monitoring device needs to be able to detect any and all possible agents. A dedicated device capable of detecting anthrax, for instance, is interesting but impractical, because it could be thwarted by the terrorist use of another agent. Also, another example is gas chromotagraphy (GC)-type systems, which though very effective for detecting volatile organics, would offer no protection against the introduction of a heavy metal agent, such as thallium.

The detection of diverse contaminants requires a realignment of thinking from the traditional development of a sensor for a given compound or agent. A definite challenge is posed in developing sensor arrays, toxicity testing, "lab on a chip" technology, or a coordinated array of analytical instrumentation that would be capable of detecting a wide variety of contaminants. Another approach is to use chemometrics to detect and characterize changes in basic water-quality parameters and correlate them with threat agent introduction.

A common misconception concerning analysis in water is that the system is stable, with little variation over time or among locations. This is probably because most analysts that are not specifically involved in water research are accustomed to using laboratory-grade deionized water as the norm when running experiments. In the real world, even after treatment, there is great diversity in source water and in the water found throughout distribution systems. For a parameter as simple as pH, which we would expect to be around 7 ± 1 pH units, the diversity is much greater than expected and can run from 3 to near 11 pH units. Also, in a given system, there is great heterogeneity over time in basic conditions including pH, turbidity, and conductivity. Figure 7–1 represents the diversity that can be found in the real world in these types of parameters at a single location over time.

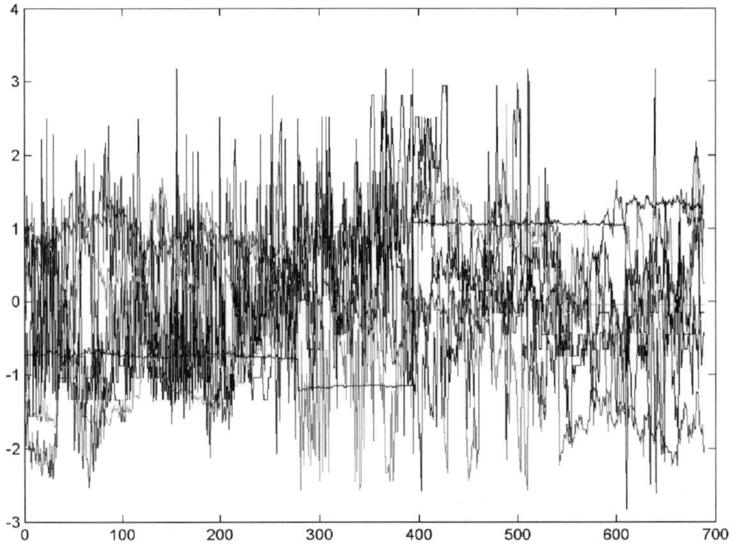

Fig. 7–1. The diversity and variability of water quality in the distribution system over time. The data are autoscaled for three months and represent pH, conductivity, residual chlorine, turbidity, and total organic carbon. This is by no means a static system. (Courtesy of Hach HST)

In addition to the great diversity of water quality in a given system, the general environment is also very harsh. The environmental conditions found in the pipes can pose a great challenge to the design of a monitoring system that is robust enough to be deployed for extended periods without becoming disabled or requiring extensive maintenance. The water conditions may be corrosive or scaling in nature. This can lead to the degradation of anything placed in the system or the formation of a coating of various types of materials (e.g., see fig. 7–2). Also present in most pipe systems is something known as biofilm. This is a thin layer of bacteria and their associated slime that coats the inside of pipes and anything else present in the system. This layer of growth can coat sensors and render them unable to function properly. It can also clog small tubes and pipes used to draw off samples, resulting in erroneous readings. Especially where source water is concerned, extreme variations in temperature and weather conditions may be encountered, as well as freeze and thaw cycles. Any detection system designed to function over long periods of time must be capable of handling these harsh conditions.

There is also the problem of aging infrastructure. Many of the water pipes in our major cities are more than one hundred years old and are occluded with rust, crumbling concrete, and other debris. Some of the pipes are actually still the original wooden pipes installed when the systems were first built. This raises concerns with regard to instillation of and sampling by a distribution system monitoring platform.

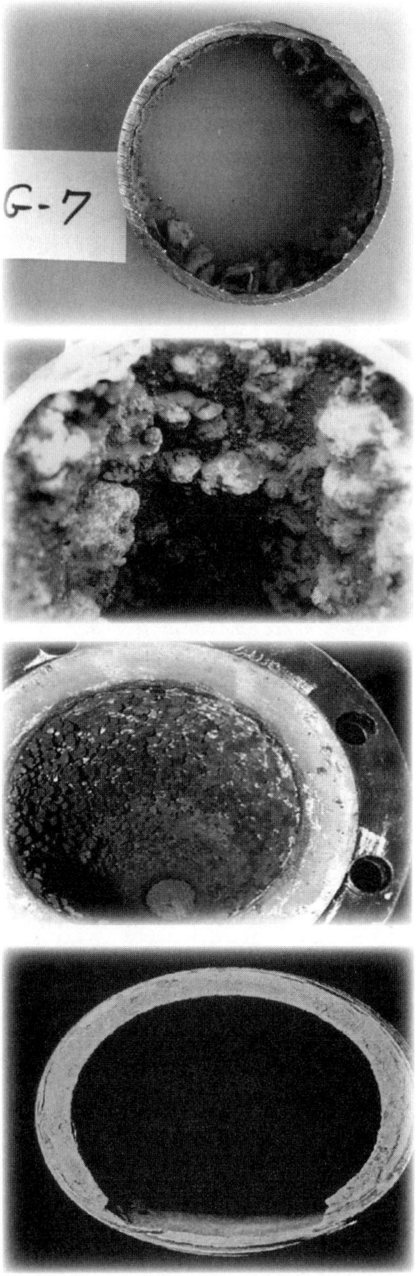

Fig. 7–2. Various forms of corrosion and scaling that could be problematic in a monitoring situation. (Photographs courtesy of the Army Corp of Engineers Research Laboratory)

Cost constraints are also a major factor considered in the design of monitoring systems. Because of the aging infrastructure that plagues most municipal water supply systems, the huge expenditure for needed upgrades leave little funding room for security. Therefore, it is very important that any sensor system be cost effective. The goal of cost-effectiveness can be addressed in two different manners. One method would be to design a very inexpensive system that could be deployed for a low per-customer cost. The other method would be to develop a system that is capable of providing data that could be useful in decreasing the day-to-day operating costs of the system and improving general water quality, thus making its cost a recoverable expense.

A smoke detector can be used as an analogy. The relatively low cost of smoke detectors allows their widespread deployment to protect against an unlikely event. If smoke detectors were to cost several thousand dollars each, few locations would be equipped with them. A water system protection device is similar. Few municipalities would fund a system, unless it were very inexpensive, that was designed to protect solely against terrorist events, because of the low likelihood of their occurrence in a given location. The market could and would bear a higher cost for a dual-use device that could streamline general operations and provide increased water quality, hence providing real value even if a terrorist event never occurs.

The remainder of this chapter will delve into these challenges presented above and present a variety of current and upcoming solutions that are being used or are being proposed to address the problems inherent in monitoring these locations.

Monitoring Source Water

As discussed in chapter 4, the vulnerability of source waters to contamination attacks is limited by a number of factors, including dilution effects and natural attenuation. Because the water also has to traverse the treatment plant barrier before it can affect consumers, the hazard is further mitigated. Even so, the risk of an intentional contamination event—in addition to the risk of an accidental spill—is real. While such an event would be unlikely to cause mass casualties, it could have a severe impact within the affected area, resulting in denial of service to that community for some time.

In November 2005, an explosion at a PetroChina factory, in the Jilin province, resulted in a massive spill of benzene into the Songhua River, in northern China. This massive spill resulted in the contamination of river water and a denial of service to over four million customers in Harbin, the capital of China's northeastern Heliongjiang province. As the contaminant plume migrated downriver and ultimately crossed the border into Russia, it affected other cities and became an international problem. This massive spill didn't result in any casualties, but it did cause a panic and the hoarding of bottled water and food in several areas.[1]

This spill was quickly reported, and protective action was taken to prevent customer exposure to the contaminated water. A scenario can be imagined in which an industrial accident is not reported or a deliberate event causes a similar incident. If the chemical or material involved were highly toxic and no monitoring was done, customers could be exposed. This emphasizes the benefits of monitoring source waters. In fact, source waters are already monitored in several locations throughout the world; the European Community has been a leader in this area for years.

Heavily industrialized areas in Europe, such as the Rhine Valley (fig. 7–3), were recognized early on as potential health hazards. In this area, heavy industrialization and a large population reliant on river water made for an accident waiting to happen. A series of industrial mishaps along the Rhine River led to the development of an early-warning system that would alert utilities downstream of an impending spill, so that water intakes could be shut down before the contaminated water reached the treatment plants. These systems employ online monitoring of a variety of physical and chemical measurements of water condition, together with various toxicological methods, to determine river water quality. This system has proven to be effective in preventing serious contamination form reaching the treatment plants.

Potential challenges in the monitoring of source water

There are a number of potential problems in implementing a monitoring system for source water. There is the problem of diurnal (daytime versus nighttime) and seasonal shifts in water quality, owing to factors such as aquatic plant respiration and decaying vegetation from autumn leaf falls. Varying amounts of sediments, turbidity, and dissolved solids, owing to precipitation events and spring runoff fluctuations,

Fig. 7–3. Dense population and heavy industrial use of the Rhine River in Germany and Holland made it a prime candidate for source water monitoring. After several accidents, a comprehensive monitoring and early warning system was implemented. (Courtesy of Degussa)

may be problematic. Monitoring equipment is often exposed to extreme conditions of heat and cold and is often located in remote areas; thus, power supply and communications become issues. Because of the transient and unexpected nature of the events we are trying to detect, the monitoring systems also need to be online and continuous; that is to say, if we knew when we needed to monitor, we wouldn't need to monitor.

One of the chief problems is what to measure. The diverse list of contaminants would make monitoring for individual chemicals or even classes into a futile effort to out-guess the terrorists. This suggests that rather than testing for each chemical or class of chemical on an individual basis some broad-spectrum form of testing should be utilized to monitor for general changes in water integrity. The most likely candidates for such monitoring would be toxicity testing and bulk parameter monitoring of traditional water quality parameters. Only online systems are considered here, as grab samples would be ineffectual as an early warning system. Other analytical methods which are not continuous real time online methodologies will be discussed in chapter 8.

Toxicity

Toxicity is the ability of a chemical or mixture of chemicals to cause adverse effects on exposure of a living organism to the compound. These effects include negative impacts on survival, growth, and reproduction. Toxicity tests are analytical experiments to detect or quantify toxicity in a sample by measuring the effects that exposure has on standard test organisms.

Toxicity testing has a long history. The food testers employed by Roman emperors and medieval kings were a rudimentary form of toxicity testing. The probability of getting a volunteer for this sort of work today is not high—and even if volunteers were available, the Occupational Safety and Health Administration (OSHA) would most likely object. Therefore, toxicity testing for threats to human health is a problem. No other organism will respond to a toxin in exactly the same way as a human. The closer an organism is to humans on the evolutionary tree, the more its responses to toxins would be expected to mirror human responses. This is why clinical toxicologists have long used other mammals, such as monkeys, dogs, rats, and mice, in toxicological studies.[2]

The expense and slow response time involved with using mammals in the monitoring of water supplies is prohibitive for general applications. The next best alternative is to use lower-order vertebrates, such as fish, as surrogates. Captive populations of fish, such as trout, have been monitored for several years, most notably in Europe, to evaluate source water for contamination before it enters treatment plants.[3]

Monitoring of fish has been deployed in a variety of settings. The simplest versions are homemade avoidance systems, in which the fish are fed in the tanks nearest to the supply of source water; as long as the quality of the water entering the tank is good, the fish will remain in that tank, but if contaminated water is introduced, the fish will migrate away to tanks progressively farther from the water inlet. This method has been used in Asia for centuries and more recently in the United States; for example, Pittsburgh employs such a system (fig. 7–4).

Fig. 7–4. Simple fish toxin avoidance setups, as the one pictured here, can be a simple means of measuring the quality of source water entering a treatment plant. (Photograph courtesy of Stanley States and the Pittsburgh Water and Sewer Authority)

Elaborate computerized instruments are also deployed in some locations, since simple avoidance of gross contamination is easily recognizable, but more elusive changes in water quality may be difficult to detect. The interpretation of changes in fish action or viability can be very subtle, and it can be difficult for unskilled observers to differentiate what is normal from what is not. Advanced technology, such as CCD cameras and advanced sensors that interpret fish movement or gill action (or even the coughing mechanism that fish undergo when under stress), have been developed to detect changes in water quality.

The system pictured in figure 7–5 was designed by the U.S. Army Center for Environmental and Health Research (USACEHR). This system monitors the electrical impulses generated during the normal respiratory actions of captive bluegills held in the instruments' tanks. As the fish move in the chamber and ventilate their gills, muscle contractions generate electrical signals in the water that are monitored by a computer. End points include ventilation rate, strength of ventilation, gill purge (cough) rate, and body movement rate. When at least six of the eight fish behave abnormally, the computer provides immediate alarm notification and starts an automated water sampler, for follow-up chemical analyses.[4]

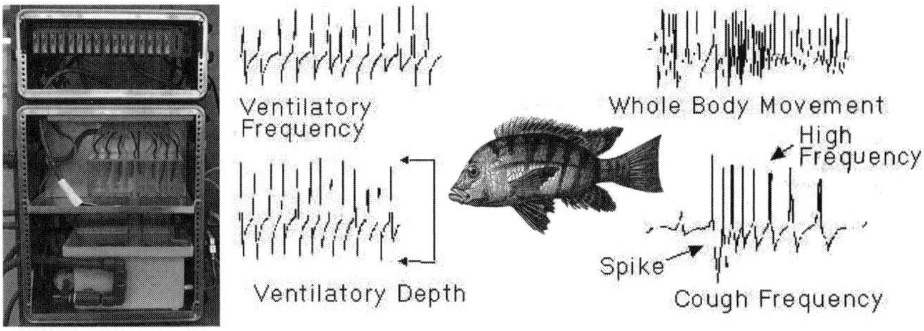

Fig. 7–5. The system designed by USACHER utilizes electrical field variations to monitor fish health.

In this type of system, interactions with parameters such as water temperature and turbidity can skew results. Many systems utilize indigenous species, such as bluegills and catfish, to lessen sensitivity to normal changes in water quality that may affect the fish. The theory is that the fish can handle these types of changes because they are adapted to the natural changes in their environment. However, one drawback to this approach is that it can at times be difficult to care for indigenous species. Also, some instrumentation requires specific size and age distributions in order to function properly. Furthermore, obtaining these stocks of indigenous species can be difficult. Many manufacturers have decided that the advantages of using indigenous species—especially the reduction of false positives—are outweighed by the hassle of dealing with them and have turned to more easily reared aquarium species. Examples are the fish biomonitors produced by Seiko Electric (the Fish-Toximeter) and Biological Monitoring (the Bio-Sensor Fish Monitor), which use the ubiquitous and easy-to-care-for guppy.

No matter what fish species is chosen, none will be perfect. Tradeoffs when choosing which species to monitor include sensitivity, false positives from natural water changes, ease of operation, and stock maintenance. These systems do, however, offer a first line of defense for evaluating the quality of feed water to treatment plants.

The usefulness of fish monitoring, unfortunately, is restricted to source water. The chlorine present after treatment would damage the fish's gill structure and would make such a system useless for monitoring of the distribution system, unless the chlorine were first removed; the removal process, however, could alter the water's toxic profile (e.g., sequestering heavy metals) in the analysis, skewing the results. Also, the wide number of points that need to be monitored in order to offer adequate surveillance of the distribution system makes fish monitoring impractical for that application. Therefore, this type of system has limited applications and would be most useful in larger systems for monitoring surface water feed supplies.

If fish rearing and maintenance appears to be too complicated an undertaking, another option is the use of invertebrates. These organisms are even further removed in terms of equivalence to human responses, but ease of use and lower maintenance costs make them a viable choice. Instruments using organisms such as the water flea *Daphnia* have been developed.[5] *Daphnia* are small mostly planktonic crustaceans, between 0.2 and 5 mm in length. They live in various aquatic environments, such as acidic swamps, lakes, ponds, streams, and rivers. Because of their sensitivity to toxins, prolific nature and ease of care, *Daphnia* have long been a standard test organism to determine the toxicity of samples.

One example of instrumentation that utilizes the water flea is the Daphnia Toximeter produced by BBE Moldaenke (fig. 7–6). The BBE Daphnia Toximeter observes *Daphnia* under the influence of a water sample that is continuously running (at a rate of 0.5 to 2 L/h) through the measuring chamber containing the *Daphnia.* Pictures obtained with a charge-coupled device (CCD) camera are analyzed online with an integrated personal computer. A change in the movement of the *Daphnia,* based on a number of different parameters, is examined; if this change is statistically significant, an alarm is induced.[6]

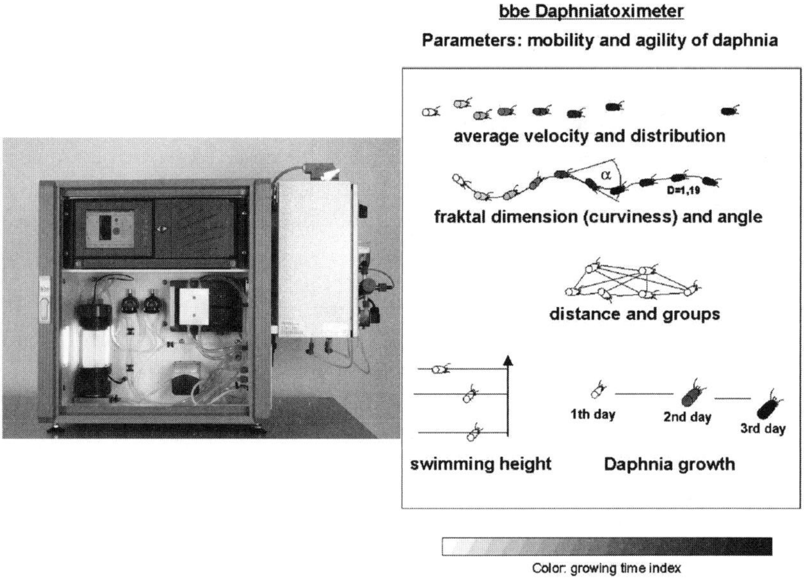

Fig. 7–6. The BBE Daphnia Toximeter utilizes a computerized system to evaluate a number of key aspects of Daphnia *behavior to determine if a toxicant is present.*

While *Daphnia* are simpler to care for than fish, instruments based on invertebrates have some of the same problems as fish-based systems, and the instruments themselves tend to be expensive. Another indicator organism that has been used is algae in the source water. All water sources that are exposed to sunlight have a naturally occurring population of various algae. This ever-present population has been utilized in a method called the WaterSentry, developed at Oak Ridge National Laboratory (ORNL) and marketed by United Defense/BAE Systems.

WaterSentry detects waterborne toxins by evaluating the natural activity of green algae. The technology is capable of detecting the presence of toxic agents at levels expected during military conflicts or terrorist attacks on civilian water supplies; it is not intended for precise chemical analysis to identify the exact agent, instead offering a general warning that conditions have become toxic to the algae. This device utilizes patented AquaSentinel technology from ORNL. Toxicity is detected by monitoring the optoelectronic fluorescence of the algae.

Fluorescence detection and monitoring is a well-understood process. By measuring the natural algae fluorescence and establishing a baseline of activity, you can detect any deviations therefrom. These deviations indicate that the algae are being stressed in some manner, perhaps by the presence of toxins in the water. The operational concept for WaterSentry is that detected deviations will alert the water supply control staff that water use from this source should be immediately stopped and a detailed chemical analysis should be performed.[7]

WaterSentry provides an effective method for tracking changes in source water resulting from a contamination event. It has been shown to respond to several compounds, including methyl parathion, cyanide, and ricin. Whether it will respond to all compounds of interest is open to question. Because it relies on indigenous algae, AquaSentinel is relegated to untreated surface water applications unless algae are introduced to the system (bringing us back to the problem of culturing and dechlorination). The instruments' use of costly fluorometric technology makes the overall price of the system fairly high. This sort of instrumentation would best be deployed near a source water intake for a plant.

Another option that is available for online toxicity monitoring is bacterial cultures. While microorganisms such as bacteria are far removed from humans and would be expected to exhibit differential responses from humans, their simplicity, low cost, and ease of use make them a desirable test organism. Microorganisms have been used to monitor toxicity for many years in the wastewater industry, to safeguard the bacterial populations in treatment systems from shutdown due to toxic exposure.

A wide variety of testing procedures and methods have been developed to monitor the toxic characteristics of influent to treatment plants. The majority of these testing procedures are test kits and have not been amended to function in an online mode. These field methods are discussed further in chapter 8.

One type of commercially available test based on bacteria that has been modified to function online measures the effect that toxins have on the light output of luminescent bacteria. The Israeli company CheckLight produces an example of this type of instrumentation, called the micro-MAC Toxscreen.

Luminous bacteria emit measurable light as a byproduct of cellular respiration. Chemophysical and biological factors that affect cellular respiration promptly alter the level of luminescence. Similarly, factors that affect the cell's integrity—especially membrane function—have a strong effect on in vivo luminescence. Hence, by simply comparing the luminescence level obtained in the suspected toxic sample with that obtained in the control (i.e., a clean water sample), one may detect very low concentrations of a broad range of toxicants. In the CheckLight version, freeze-dried cultures and buffers specific for organic and inorganic contaminants enhance sensitivity and ease of use.[8]

This type of test correlates well with total toxicity, but it does have limitations. Bioluminescence is not an essential metabolic pathway, nor is it widespread among living organisms. Toxins that specifically inhibit luciferase, the enzyme responsible for bioluminescence, may not exhibit general toxicity to other organisms. In addition, the measurement of bacterial luminescence requires the use of very expensive instrumentation known as a luminometer.[2]

On the whole, the use of toxicity measurements to safeguard source water has significant merit. These systems have the ability to detect a wide variety of potential contaminants. They offer a good first line of defense to prevent contaminated source water from entering treatment plants. In many cases, however, toxicity measurement systems may not be the ideal solution because of drawbacks such as cost, false alarms from normal background occurrences not related to contamination, their unproven ability to alarm on all possible contaminants, and maintenance and culturing problems. A lower-cost, less-hassle solution may be required in some deployment scenarios, so that more sites can be monitored.

Bulk parameter monitoring

Another option for the monitoring of source water is bulk parameter monitoring. The concept involves actively and continuously monitoring basic water-quality parameters, looking for significant changes that may be indicative of a terrorist event. A variety of source water parameters can be applied to this sort of system. For many years, manufacturers in the environmental market have combined this sort of instrumentation into self-contained data collection bundles that can feed data back to a central location via wireless telemetry or hardwired packages (fig. 7–7). A number of parameters are currently available for online monitoring (see table 7–1).

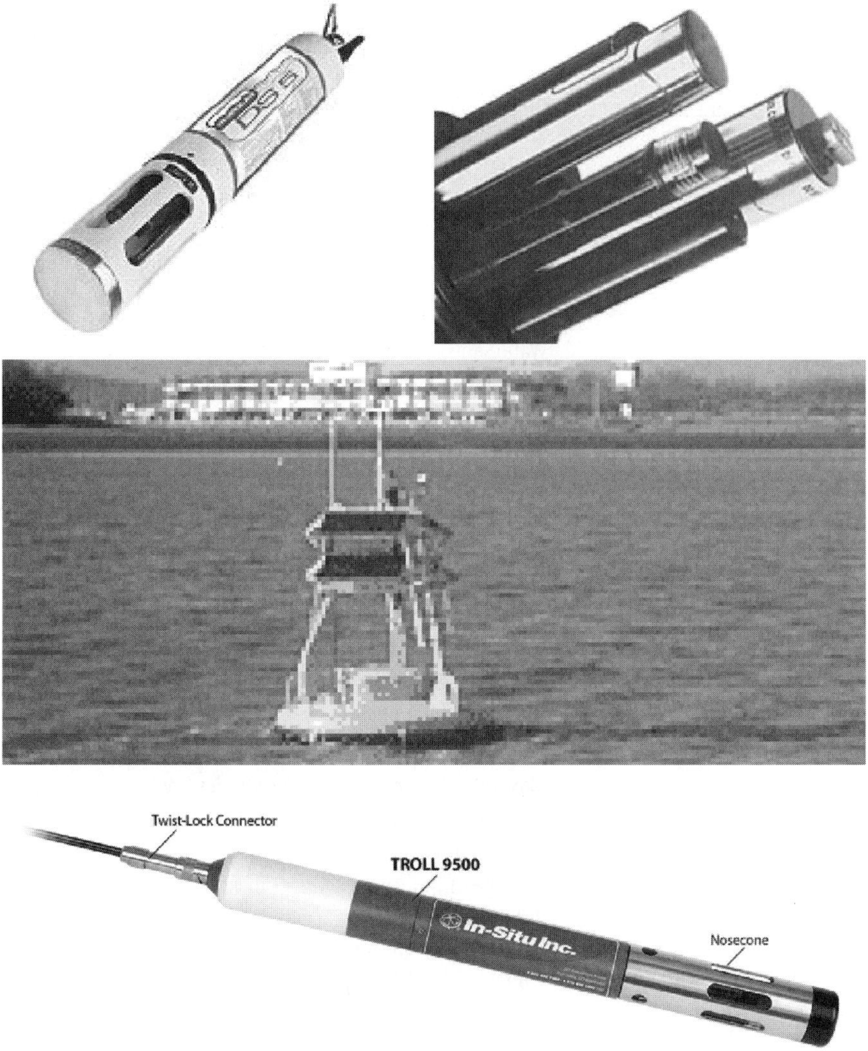

Fig. 7–7. Multiparameter probes can be deployed to monitor a variety of conditions in source water. The customer can usually choose which parameters to incorporate and how to deploy. (Photographs courtesy of HydroLab, a Hach company, and In-Situ Inc.)

Table 7–1. Parameters available for online monitoring of source water

Parameter	Applicability to Security Monitoring
Ammonium/Ammonia	May help in detecting byproducts from increased bacterial counts such as a sewage spill.
Blue-green algae	Changes can indicate toxicity to the algae as in the WaterSentry discussed previously.
Chloride	May indicate presence of chloride containing metal complexes.
Chlorophyll A	Changes can indicate toxicity to the algae as in the WaterSentry discussed previously.
Conductivity/TDS	May indicate presence of ionic species.
Dissolved oxygen (optical or polarographic)	Sudden change may indicate toxic conditions that effect algal respiration or increased levels of bacteria using up the oxygen.
Dissolved gasses (total)	Changes can correlate with some compounds.
Light ambient	May indicate an opaque plume of a toxic compound.
Nitrate	May help in detecting byproducts from increased bacterial counts such as a sewage spill.
Oxidation reduction potential (ORP)	May indicate sudden changes for oxidative or reducing species introduced into the water.
pH	Acid–base relationships
Rhodamine	Tracer used in studies to trace plumes.
Salinity	May indicate presence of ionic species.
Turbidity	May indicate some chemical compounds or increased bacterial levels.
UV absorption	Capable of indicating changes in concentration of some organic species.
General physical parameters (barometric pressure, depth, temperature etc.)	Little security utilization but can help coordinate and adjust readings from multiple locations.

Most manufacturers offer the customer a choice of which parameters to monitor. The majority of these sensors employ traditional electrochemistry methods for the parameters in question. Exceptions are optical measurements of turbidity and the new luminescent quenching methods offered for dissolved oxygen. Also, the chlorophyll and algae methods are based on UV fluorescence. The relatively low cost of these instrumentation packages allows monitoring at a variety of sites, but a huge amount of data must be analyzed and correlated to determine whether a change is significant, rather than the result of natural variation. Various programs are underway to determine whether it is feasible to transmit these data via satellite to central mega-computers that would interpret the data and look for significant changes. Intelligent PC-based algorithms are also under investigation and are undergoing field trials.

Another approach is the use of UV absorption. The Spectro::Lyser, an instrument produced by Messtechnik GmbH of Austria, uses UV absorption at a variety of wavelengths to detect organic contaminants. Algorithms interpret the incoming absorption spectrums to determine when a contaminant is present. The system can detect organics that have an absorption signal in the UV range. The system allows up to eight different alarm parameters to be set. The identification of a single substance or a group of substances is limited to those that are detectable in the UV spectrum and were implemented in the setup procedure. While the Spectro::Lyser can detect phenols, benzene, toluene, xylene, some pesticides, some nerve gases, oils, and many other substances, it would be ineffective in detecting short-chain aliphatic compounds or inorganic compounds, such as metals.[9]

Nevertheless, there are problems with bulk parameter monitoring in source water. Because of the location of deployment, there is a tendency for the sensors to become rapidly fouled, owing to biofilm buildup. A variety of remediation techniques have been deployed to lessen this problem, including automatic wipers and high-pressure bubble jets for removing the film. All of these methods help, but fouling can persist, leading to false readings, which may trigger an alarm. Even when antifouling tools are deployed, it is a good idea to have a regular maintenance program to check for and remove any fouling.

The Distribution System

As discussed previously, the distribution system represents the largest vulnerability of drinking water supplies. The risks of a deliberate contamination attack on the distribution system are high, and the consequences could be catastrophic. The only viable protection mechanism to guard against such an attack is monitoring. Dirty and occluded pipes, large geographic distribution, poorly understood hydraulics, variable water quality, multiple sources feeding a single pipe, the presence of treatment chemicals, a lack of funding, and where to locate monitors are some of the problems to be overcome when monitoring in the distribution system is considered.

The science of continual monitoring in the distribution system is giving rise to a new discipline. The distribution system is considered by many in the industry to be the last frontier in water-quality monitoring. Currently, other than occasional grab samples, monitoring in the distribution system is not regulated or performed extensively, quite possibly owing to the lack of adequate methodology. The events of 9/11 put the problem into new perspective and led to the devotion of money and research to address the problems inherent in such an undertaking. Various methods and technologies have been deployed to develop an integrated approach to monitoring that could help lessen the risks associated with contamination events, whether terror related or accidental.

Toxicity

As was discussed in the previous section on source water monitoring, toxicity testing appears to be a reasonable choice when attempting to detect the wide variety of potential threat agents that exist. There have been attempts to deploy similar online techniques to monitoring of the distribution system. As a rule, distribution water tends to be less variable than that found in source water, which should make the monitoring of toxicity easier by reducing the numbers of false positives; however, there are a number of problems that are inherent to the use of toxicity methods in the distribution system.

The main problem is culturing and maintenance. Because of the labyrinthine nature of the distribution system and because it can be accessed at virtually any point in the system, the deployment of monitoring platforms is required in a large number of locations in order to achieve an adequate level of protection for a given system. The use of live organisms in toxicity-monitoring systems requires a lot of hands-on time to produce and maintain the organisms. Even systems with automated care and feeding need an inordinate amount of looking after. These are not factors that are amenable to a widely distributed network of sensors.

Another factor that makes toxicity monitoring in the distribution system a questionable undertaking is that many of the treatment chemicals that are present in the water once it has made it to the distribution system are in themselves toxic to the organisms used. Chlorine, chloramines, fluoride, and other treatment chemicals can affect the response of the test organisms, if they do not outright kill them. For a toxicity-monitoring system to function appropriately, chemicals such as chlorine must be removed from he water before the organisms are exposed to it. Chlorine can be easily removed by treatment with dechlorinating agents such as thiosulfate. However, the process of removing the chlorine can alter the toxicity of the water. For example, if a system that has been contaminated with a mercury compound is dechlorinated with sodium thiosulfate, binding to the thiosulfate (for which it has a very great affinity) will sequester the mercury. This binding to thiosulfate can render the mercury unavailable, hence altering the toxicity profile of the water.

Another drawback to toxicity monitoring in the distribution system is the variable environment that can come into play in the network. Some organisms, such as fish, can be quite sensitive to changes in the general surroundings. Increases in vibrations and noise levels at times of peak traffic could become problematic and lead to false alarms. Shielding organisms from this type of upset can be costly or can severely limit your options for deployment. In the distribution system, where widespread deployment is a must, toxicity monitoring is not a truly viable option; rather, it plays a more useful role in the monitoring of source water, where monitoring points are fewer and more easily controlled.

Lab-on-a-chip technologies

Microelectromechanical systems (MEMS), also known as lab-on-a-chip technologies, are an innovation that is rapidly growing and finding use in the medical technology field. These are basically microscale devices that miniaturize and streamline traditional analytical and biochemical methods for mass production, facilitated by many of the same fabrication techniques as the computer chip and microelectronics industry. Mass production techniques allow the manufacture of very low-cost devices as compared to traditional instrumentation. As the technology becomes more robust, it has branched out from traditional medical diagnostics, finding use in other analytical fields. Water is no exception, and copious research is being done to amend these types of devices to water analysis.

One project currently under development is the Micro Bio Chem Lab, produced by the Sandia National Laboratory.[10] This device uses microfluidics and microchemical techniques to sample and analyze water for a variety of components, including harmful bacteria and viruses. These devices are miniaturized discrete analyzers that test for specific substances. They can use a variety of detection techniques, such as miniaturized GC or gel electrophoresis systems. Recently, projects have been initiated to adapt this technology from an independent handheld device to an online configuration, for monitoring of water supplies.

Whether these efforts to bring the system into the realm of online distribution monitoring will be successful remains to be seen. If they are successful, the low cost of the instrumentation could allow monitoring at a huge number of sites in a cost-effective manner. However, several problems with these types of systems must be overcome before they can be successful as water analyzers.

These devices rely on microfluidic techniques to draw and analyze samples. The distribution system, by its very nature, is not a friendly environment for such techniques. The aging distribution infrastructure is plagued with particulate matter. Rust particles and debris are common components of the water in these systems. These particles could easily block the micro channels in these devices. Attempts to prefilter the sample could alter the toxicity characteristics. Another problem is that, as currently designed and deployed, these devices are discrete analyzers that detect specific toxins or classes of toxics; therefore, they could be thwarted by use of a toxic material that the instrumentation was not designed to detect.

Gas chromatography

GC methods have been modified as online tools that work in a batch mode. GC can be used to separate organic compounds that are volatile. GC consists of a flowing gas (mobile) phase, an injection port, a separation column containing the stationary phase, a detector, and a data-recording system. The organic compounds are separated according to differences in their partitioning behavior between the mobile phase and the stationary phase.

The largest drawback to this technique is the limited scope of compounds that are detected. Only volatile organics are amenable to being analyzed by this method. Also, some of the instrumentation can be touchy, and the cost per deployed unit can be quite high.

Optical methods

One example of optical methods being utilized to address the problem of monitoring in the distribution system is a device called the BioSentry, produced by JMAR Technologies. The BioSentry utilizes laser-produced, multi-angle light-scattering (MALS) technology to generate unique bio-optical signatures for each microorganism. Similar to a laser turbidimeter, the BioSentry uses lasers to interrogate a water sample and analyze how a particle in the water refracts the light.

The difference with MALS technology lies in the multi-angle aspect. Rather than just reading at 90°, a number of different angles are monitored at once. This allows the generation of a three-dimensional pattern that represents the structure and size of the particle in the laser's path. This creates a pattern that is representative of the particle in question, in the same way that a shadow of an object gives some indication as to its size and shape. This pattern can then be compared to a library of patterns for different types of organisms and classified using JMAR's pathogen-detection library.

This appears to be an effective method for monitoring water for biological contamination. However, it would be ineffective against chemical contaminants. Another potential drawback is sample size. Because of the very small path capable of being monitored by the laser, only a small sample can be analyzed. Thus, it would be easy to miss bacterial or protozoan contaminants that were present at very low concentrations. This instrumentation would be able to sense large contamination events, but chances are that low-level contamination events would be missed.

Bulk parameter monitoring

Bulk parameter monitoring is the method of monitoring common water quality parameters and then looking for anomalies that may be indicative of a water contamination event. Immediately after 9/11, the applicability of deploying common sensors for water security was investigated. A number of government,[12] academic,[13] and private industry[14] studies evaluated a variety of sensors to see whether they would respond to the contaminants most likely to be used by a terrorist in an attack.

At Hach HST in early 2002, the possibility of using well characterized off-the-shelf sensors that had been proven robust for field deployment in a multi-parameter array was investigated. The sensors chosen for investigation were pH, conductivity, chlorine residual, turbidity, and total organic carbon (TOC). Data were collected for oxidation-reduction potential (ORP) but were not used in the final system, because the probes are unstable and prone to fouling in long-term installations.

As an initial investigation, the possible threat agents were divided into several broad categories. A number of agents were selected for availability and as good representatives of wider classes of possible threat agents. It was decided that each agent would be tested at three different concentrations with a test duration of one hour. The levels tested were at or below an LD_{50}, defined as the amount of agent that would be fatal to 50% of the adults exposed by drinking one liter of the contaminated water in a single day. The following agents were tested:

- Thallium: Representative of the possible metal-based agents that could be used
- Cyanide: Toxic industrial chemical
- Methomyl: Representative of possible agricultural insecticides that could be used
- Aflatoxin β_1: Representative of the biotoxins that could be used; similar to T-2 toxin
- Ethoprophos: Used as a surrogate for cholinesterase-inhibiting nerve agents such as VX
- *E. coli*: Representative of possible bacterial agents

During the test, 1.5 liters of tap water was placed in a 2-liter beaker. Residual chlorine levels in the local tap water at the Hach facility in Loveland, Colorado, are quite low, at <0.2 mg/L. Therefore, additional chlorine was added to get the levels closer to 1 mg/L. The water was then stirred for 30–40 minutes to reach equilibrium with the room air, so that any shifts in pH, TOC, and so forth were due to the agent spikes and not ambient water conditions such as an increase in dissolved CO_2. The proper amount of agent to be tested was spiked into the beaker, and readings were taken after 1 minute and every 6 minutes thereafter. The parameters listed previously were measured for a time duration of one hour. All of the tested agents showed a distinct change in one or more of the measured parameters. As an example, data for *E. coli* are presented in figure 7–8.

All of the classes of agents studied showed a significant response to at least one of the parameters measured. The response of each class of agent, when scrutinized across the entire matrix of measured parameters, showed marked differences from other classes of reagents (table 7–2). This was even more unique when speed of response and various ratios are calculated to give a unique fingerprint to each type of compound.

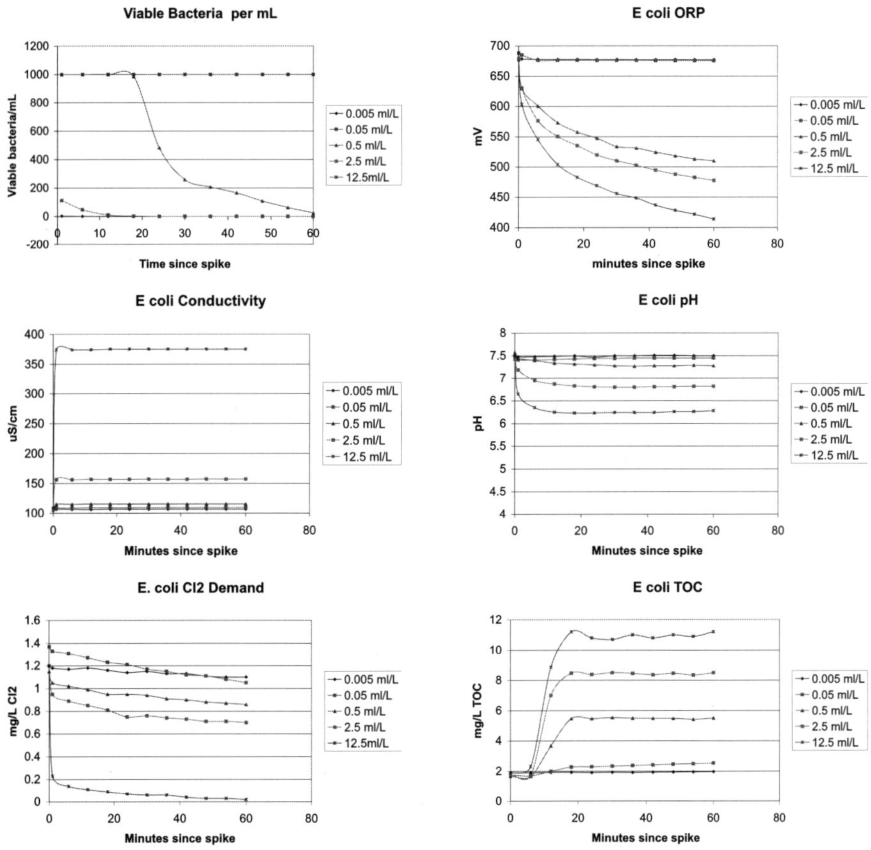

Fig. 7–8. The responses of various common water sensors as differing levels of E. coli *culture are injected into the system*

Table 7–2. Different patterns of response

Compound	Class	Low Conc. mg/L	TOC	Cl2	pH	ORP	Cond.	TDS	Fingerprint (actual algorithm not used)
Thallium	Metals	500	NT	NT	-	0	+	+	NT
E. coli	Bacteria	0.005	0/+	-	0/+	0/–	0/+	0/+	–0.700
Methomyl	Insecticide	10	+	--	-	0	0	0	–1.000
Aflatoxin	Mycotoxin	0.67	+	-	-	0	0	0	–1.944
NaCN	Industrial	50	+	-	-	+	+	+	–2.761
Ethoprophos	Nerve Agent	1	0/+	0/-	-	0	0	0	–3.000

	Strong Positive response	0/–	Weak Negative Response
0/+	Weak Positive Response	NT	Not Tested
–	Strong negative Response		

The current state of online monitoring with existing instrumentation is such that significant actual events should be detectable. The problem, though, is what to do with all the data. Enormous amounts of streaming data need to be processed. Another problem is the minute-to-minute variability that is present in a system. How are we to determine whether alterations in water-quality parameters are significant against a background of dynamic changes? In the words of Russell Young, director of advanced technology development at Hach, "How can we see the candle against the sun?"

Unless a full-time team of statisticians is to be employed to make sense of this information, intelligent algorithms will be necessary, to streamline the process. Intelligent algorithms should be capable of detecting the subtle changes in bulk parameter readings that are indicative of an incursion into the system. They should also be capable of differentiating the unique pattern of responses that are elicited by different classes of agent. These differences may be enough to narrow down the cause of events and, possibly, to fingerprint the class, if not the most likely members of that class, that caused an event. One such approach is detailed in the following paragraphs.

A system designed by Hach HST makes use of five common bulk parameters that are monitored simultaneously in real time. The parameters that are monitored are pH, conductivity, TOC, turbidity, and residual chlorine. When measured in real time, these parameters can show a lot of variability in a given system. That is why it is must for the design of such a system to incorporate a baseline estimator that is sensitive to small perturbations and yet resilient enough to not be constantly raising alarms triggered by normal fluctuations. Many classical methods of baseline determination result in poor sensitivity or high rates of false alarms. The proprietary baseline estimator used in this system addresses these problems (see fig. 7–9).

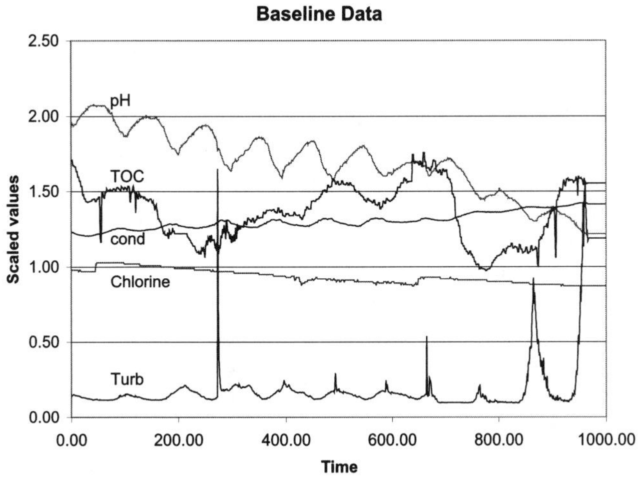

Fig. 7–9. Real-world baseline data. The variability in bulk water parameters that is common in the distribution system requires that any algorithm contain a workable baseline estimator.

In the system as it is designed, the signals from all of the instruments are processed from a five-paramater measure into a single scalar trigger signal in an event monitor computer system that contains the algorithm. The signal then goes through the crucial proprietary baseline estimator. A deviation of the signal from the estimated baseline is then derived. Then, a gain matrix is applied, to weight the various parameters on the basis of experimental data for a wide variety of probable threat agents. The magnitude of the deviation signal is then compared to a preset threshold level. If the signal exceeds the threshold, the trigger is activated. Figure 7–10 shows the same data processed through the algorithm.

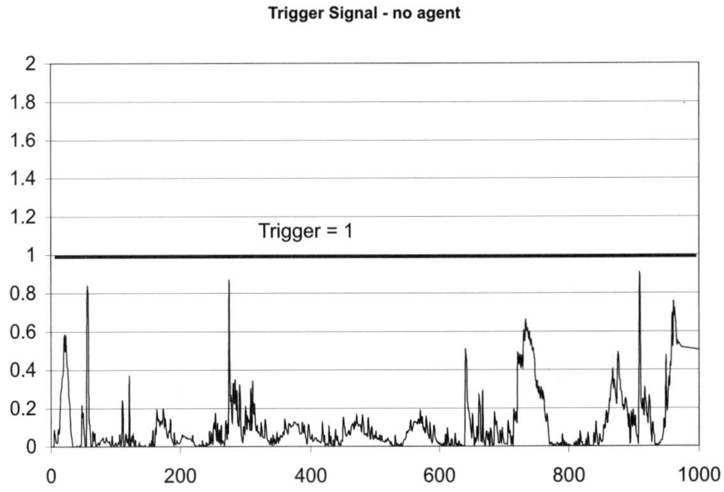

Fig. 7–10. The noisy data from figure 7–9 become easy to interpret when processed through an algorithm. In this case, no significant events above a threshold of 1 are occurring; therefore, no trigger is initiated.

Even with extremely noisy data, the system does not trigger at a threshold level set at 1. Therefore, during normal operation, with no agent present, the process deviation should not be large enough to produce a trigger signal >1. However, when the data for a cyanide incursion at 1% of the LD_{50} (or approximately 2.8 mg/L) is superimposed on the system, the trigger level of 1 is easily exceeded (see fig. 7–11). Other contaminants exhibit similar results.

The unknown alarm rate when the system is tracking real-world data is also quite low. The system is equipped with a learning algorithm, so that as unknown alarm events occur over time, the system has the ability to store the signature generated during the event. The operator can then go into the program and identify that function and associate it with a known cause, such as turning on a pump or switching water sources. Then, the next time that the event occurs, it will be recognized and identified appropriately. As the system learns, the probability of an alarm that has not been previously encountered and thus remains unidentified will decrease and eventually approach zero. The probability of an unknown alarm due to a given event depends on the frequency of the occurrence of such an event and the time that the algorithm has had to learn that event. Events that occur frequently will be quickly learned, while rare or singular events will take longer to be learned and stored. This should result in a fairly rapid decrease in the number of unknown alarms, because common events are quickly learned.

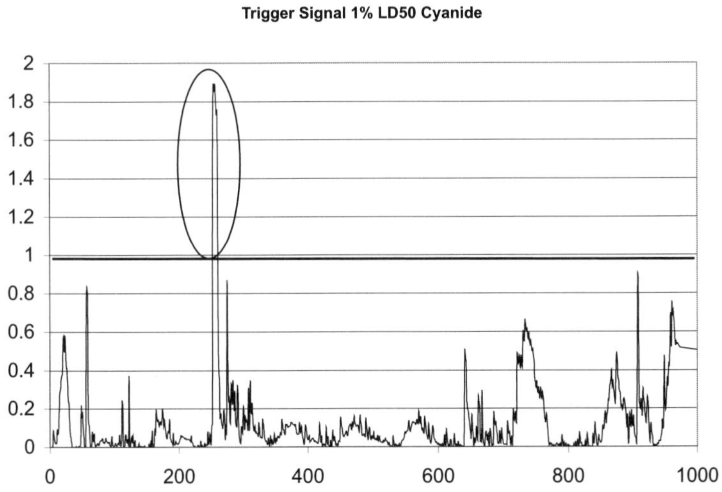

Fig. 7–11. The ability of the algorithm system to differentiate and trigger based on low levels of contaminants against a noisy background

The deviation vector that is derived from the trigger algorithm contains significantly more data than needed to simply trigger the system. The deviation vector's magnitude relates to concentration and trigger signal, while the deviation vector direction relates to the agent characteristics. Therefore, laboratory agent data can be used to build a threat agent library of deviation vectors. A deviation vector from the water monitor can be compared to agent vectors in the threat agent library to determine whether there is a match within a tolerance. This system can be used to classify the agent(s) present.

Each vector results in a vector angle in n-space that, based on the research conducted so far, appears to be unique. Figure 7–12 is a radar plot representation of agent data that visually illustrates this point. Because the direction of the vector is unique for a given agent, the algorithm can be used to classify the cause of a trigger. When the event trigger is set off, the library search begins. The agent library is given priority and is searched first. If a match is made, the agent is identified. If no match is found, the plant library is then searched, and the event is identified if it matches one of the vectors in the plant library. If no match is found, the data are saved, and the operator can enter an ID when a match is determined. The agent library is provided with the system, and the plant library is learned on site. If an attack occurs somewhere else in the country, the vector is saved and can be downloaded to any other plant that has the system. The agent library may be updated as new profiles are generated.

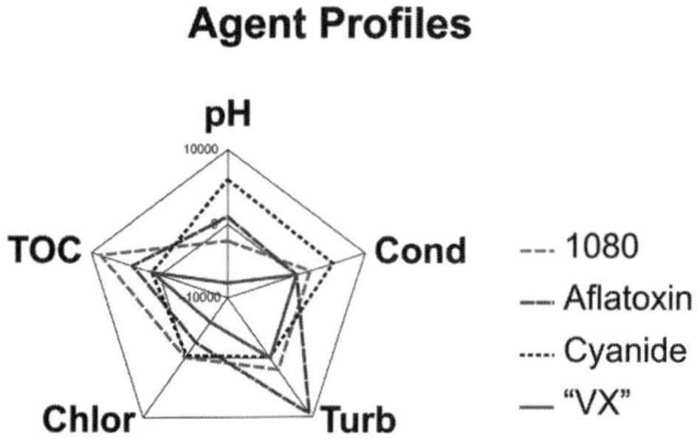

Fig. 7–12. Radar plot demonstrating how the different response characteristics of individual threat agents leads to unique vector directions in 5-space that can be used to classify events

There has been some conjecture from academics, among others, that building a threat agent database is straightforward. There are ongoing attempts to gather such data without conducting actual experiments, instead using computational methods. Research at Hach HST has shown this not to be a viable method of fingerprinting. Many unexpected reactions and responses occur in the actual experiments.

Building the threat agent library is not trivial and requires extensive laboratory work. First, it is necessary to run laboratory experiments in several different matrixes to derive instrument response to the agents. Then, it is necessary to mathematically examine the magnitude of response and unit vector structure to see if it will work using the method. The next step is to simulate addition of the agent to real water data at different dose levels, to determine the ability to trigger on and to detect/classify the agent. At this point, it is possible to add the agent to the library.

There has also been conjecture that simply monitoring pH and chlorine levels may be adequate for a security system. Experiments have shown that a significant number of compounds that are threat agents that would be expected to respond to a change in chlorine levels in fact do not. This may be because of complete unreactivity or extremely slow kinetics of the reaction between chlorine and the compound. This stresses the importance of collection of experimental data and addition of supplemental functions (above and beyond chlorine and pH) to a monitoring system; others parameters, such as total organic carbon are needed, for enhancement of the measurements.

Notably, however, as soon as the system is turned on, it will be actively working and will have the ability to trigger and classify immediately if the signature of a known threat agent is encountered. But even more uniquely and importantly, if a completely unknown agent is introduced at levels that exceed the threshold signal level, the system will trigger and classify it as an unknown that warrants further investigation. This attribute of the system is unique and is the only available method of which I am aware that will respond to and trigger an alarm based on totally unknown contaminants.

A number of similar multiparameter measurement platforms without the addition of intelligent algorithms have been evaluated for such applications.[15] These systems appear to be a good choice for detecting water-quality excursions that could be linked to water security events. There are a number of advantages to using such systems. The chief advantage is that these instruments are not new. They are common everyday parameters with which the average industry worker is quite familiar, thus adding a degree of comfort in operations that is not afforded by other new technology. As existing technologies, these instruments have been proven to be robust and dependable in prior field deployments. They give measurements that would be of practical interest to water utility personnel, above and beyond their use as water security devices.

For example, through many years of experience, the best personnel at treatment plant operations have developed a sense for knowing if something in the treatment system is amiss. It can be a smell, a color, clarity (or lack thereof), a sound, or just a tingling in the nape of the neck. One gains this sense only by extensive experience in a particular facility. I contend that such senses do not exist in distribution systems because there has typically been little measurement done through which to gain these senses. Therefore, bulk parameter monitoring in the distribution system with interpretive algorithms has the potential to become the artificial sense that quickly learns the quirks of the distribution system; further those quirks can then be labeled by those with extensive experience, so that less experienced employees have the benefit of that knowledge without having to wait 5 or 10 years or even longer.

A good phrase to describe this knowledge base would be "institutional intuition."[16] With the aging of the work force and rapid employee turnover, institutional intuition could quickly die out. Algorithms could circumvent this loss of knowledge and build a knowledge base that has not previously existed. This could allow improvements in system operation that may result in cost savings and definitely will result in the delivery of a higher-quality product to the consumer

One key advantage of the multiparameter array is the ability to detect such a wide variety of potential threat agents, from metals to organics and biological agents. The ability to trigger based on unique unknown events is also a major plus.

Some disadvantages are that events that occur during normal operation may trigger an unknown alarm. This, however, can be an advantage if the information is used to generate institutional intuition. Nonetheless, this learning phase is not free and requires an input of time and effort, to investigate and classify these alarms, so that they can be placed into the database.

Another disadvantage of such systems is that, while they will detect biological events, they are not as sensitive to such events as are other methods. The majority of the detection capability comes from the growth media introduced along with the biological agents. They do not perform cell counts, nor do they carry out individual bacterial identification. One biological event tends to look pretty much like another. Such systems are not likely to pick up on a very low level of bacteria in the system, and for the EPA requirement of <1 coliforms per 100 mL sample, they would not respond. However, as discussed previously (in chap. 4), the most likely attack with a biological agent would include media, both for ease of use and to degrade chlorine levels to a level at which the bacteria could survive. In these cases, the instruments would respond.

A further problem has to do with deployment. Many of these instrument packages tend to be somewhat large and require a suitable site for deployment. Many also generate a waste stream that needs to be dealt with. These size and waste constraints can limit where these types of systems can be deployed. There are, however, options for other means of measuring these parameters than traditional wet chemistry and optics. These include electrochemical and microscale devices that can be inserted directly into pipes (see fig. 7–13). Microchemical-based devices tend to suffer from problems with robustness, as detailed in the section on lab-on-a-chip technology. These and other electrochemical methods offer less sensitivity than traditional means of measuring bulk parameters. They may be more constrained as to what water conditions they require for proper functioning (e.g., electrochemical chlorine measurement may be effective only in a limited pH range); however, they may be the only option for some deployment scenarios.

How and where to deploy

Each water distribution system is unique in its configuration and hydraulics, and each community is unique in its priorities of where and what to protect first. Because of cost constraints and technical proscriptions on where systems can be deployed, it is virtually impossible to 100% protect every tap. The sentiment of some water system managers is that if they cannot protect everyone, then they will just not make the choice at all; they don't want to be held responsible if anything happens and casualties result. "Why wasn't everyone protected?" is a headline that they prefer to avoid at all costs.

This mode of operations is short sighted. Any decrease in casualties has to be looked on as a plus. Furthermore, failure to take action of some sort, no matter how limited, may open up the utilities to a larger liability than deploying existing technology to the best of their ability.

Fig. 7–13. At times, more sensitive traditional methods of bulk parameter monitoring are not acceptable, owing to size and waste constraints. In-pipe deployment models are available but may suffer from decreased sensitivity or robustness. Pictured in the upper panel is the Hach Water Security Distribution Monitoring Panel, employing traditional monitoring technology; in the lower left is the Hach PipeSonde, an in-pipe electrochemical and miniaturized optical package; in the lower right is the DasCore Six-CENSE microchip-based in-pipe probe.

After the choice has been made to monitor, the decision as to what type of monitoring system to deploy must be made. The wide variety of types of systems and their different capabilities makes this a complex problem. Each system has advantages and disadvantages.

The choice of what to deploy is not necessarily an either-or decision. The best choice may be a network configuration that deploys different types and cost ranges of sensors in different areas, to give optimum coverage and capabilities (fig. 7–14). Although not every point will receive complete protection, a network approach has the best chance of detecting an event early in its onset and of alerting the system operators, so that they can make the crucial decisions that will be needed to limit the damage. If an attack is detected early, consumers can be warned not to use the water. Also, turning off valves can make possible the isolation of the contaminant plume to a small area before the entire system becomes unusable.

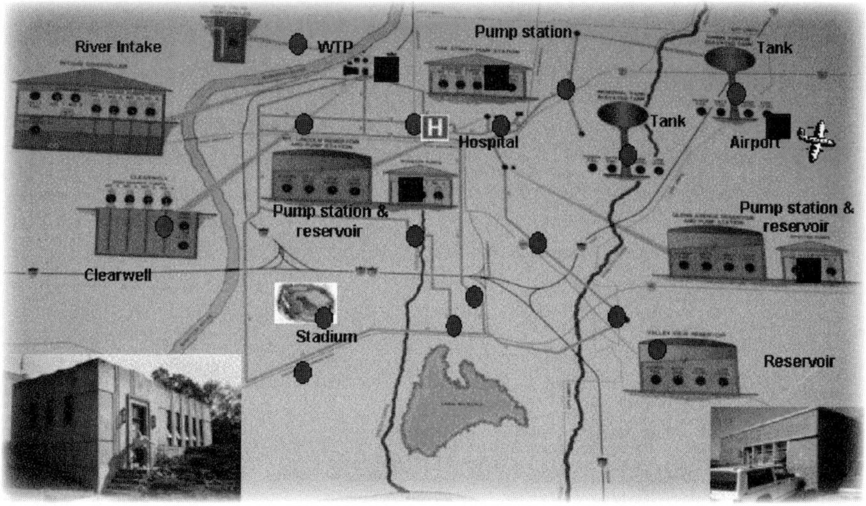

Fig. 7–14. Network deployment strategy. Squares represent possible deployment sites for a more sophisticated and expensive platform; circles represent possible deployment sites for a midpriced, less-capable solution, to be used in conjunction with the higher-priced systems.

After the choice has been made as to what type of monitoring system to deploy, the problem remains of where best to locate the systems in the distribution network. While this is not any easy decision to make, it is a necessary one.

The first step is to determine what are the key areas to protect. Likely targets and critical assets should be identified. Many of the initial vulnerability assessments completed after 9/11 failed to adequately address the distribution system. As the severe vulnerability of this area has become clear, managers should not only address the distribution system as a whole but break it down into critical areas in need of protection. These could include such facilities as schools, hospitals, large sporting and entertainment venues, military facilities, government (political) and icon facilities, large office or apartment buildings, areas of extremely dense population, and so forth. After decisions have been made as to what areas absolutely must be protected, questions remains as to exactly where to place the sensors for optimum coverage and, when expanding an existing system of monitors to attain more general coverage, where they should be placed to be most cost-effective.

To answer these questions, the EPA's National Homeland Security Research Center, Water Infrastructure Protection Division, has initiated a program called the Threat Ensemble Vulnerability Assessment (TEVA) program. This program's goals are to thoroughly study contamination threats to drinking water systems and use the information gained to design monitoring and surveillance systems and other mitigation methods to prepare for and respond to contamination attacks on drinking water systems.

TEVA uses the EPANET hydraulic water-quality model, as well as the multi-species modules for EPANET, to simulate the fate and transport of contaminants in distribution systems. By considering the uncertainty of potential contamination scenarios, TEVA calculates the statistical distribution of potential health impacts. Consequences of the contamination are estimated by predicting the public health impacts resulting from the ingestion of contaminated water. Using a probabilistic model for ingestion, contaminant-specific dose-response models, and dynamic models for disease progression over time, TEVA can predict health impacts.

TEVA focuses on the design of contaminant warning systems to mitigate the effects of contamination events. Contaminant warning systems collect information from online sensors to provide an early warning of a contamination event and to reduce public health or economic impacts. The TEVA framework offers several options for optimally locating sensors and allows for the comparison of costs and benefits of various sensor-network designs.

The TEVA computational framework is intended for use in other scenarios, besides terrorist attacks. The software will model accidental contamination resulting from backflow, as well as cross-connection and permeation or leaching from system components, and can design monitoring networks to detect such contaminants. A preliminary version of the software is expected to be available for research purposes by the end of 2006.[17]

Syndromic Surveillance

Direct monitoring of water quality is not the only approach that is being investigated to provide detection and warning of an attack. Copious research dollars are being spent to investigate the utilization of syndromic surveillance as an alternative methodology. Syndromic surveillance is a concept borrowed from the medical profession. In medical usage, the term refers to surveillance using health-related data that precede diagnosis and signal a sufficient probability of a case or an outbreak to warrant further public health response. Although syndromic surveillance has historically been utilized to target disease outbreaks, its utility for detecting outbreaks associated with bioterrorism is increasingly being explored by public health officials.

In the homeland security usage—and as it pertains to protecting water—syndromic surveillance is the use of advanced computational techniques and data-mining algorithms to monitor a number of nonspecific indicators of a possible attack. These include such data as hospital admissions, 911 calls, pharmacy sales, and complaints to the utility. These data streams are directed to a centralized computing system that correlates all of the factors and extrapolates the probability of an attack by using advanced algorithms. Once a probable attack has been indicated, appropriate response actions can be initiated to treat the potential victims.

Even though much useful information could theoretically be extrapolated from such a monitoring program, there are severe drawbacks. Syndromic surveillance, by definition, is directed toward thwarting naturally occurring outbreaks of disease. An intentional contamination event using water as a vector may spread quickly enough to make detection by such a mode redundant and unnecessary. By the time multiple victims show up at the hospital, it may be obvious that an attack is underway. Also, the reliance on such a mode of detection delays the reporting of the hypothetical event until actual exposures have occurred. This may be adequate in cases of a bacterial contaminant that has a fairly long incubation period and can be treated with antibiotics; it is, however, woefully inadequate in the case of a chemical or biotoxin contamination event.

Many chemicals and biotoxins are not detectable by the consumer; because they have no taste or odor, customer complaints must be eliminated as a viable data source. Also, the onset of symptoms after exposure may be delayed, as is the case with many bacterial contaminants. The problem is that many chemical contaminants have no known treatment after exposure. If you ingest enough, you are plain and simply going to die.

In these cases, the reliance on hospital admissions, pharmaceutical sales, and so forth becomes nothing more than a body count technology. By the time such an attack is detected via syndromic surveillance, it is too late to do anything to decrease the number of casualties and the damage incurred. Ironically, under federal Superfund statutes, if industry were to use such means to monitor for public safety, such an approach would be viewed as illegal, the equivalent of using the public as guinea pigs.

The use of such technology as a stand-alone method becomes nothing more than a means to track damage, rather than to prevent it. Syndromic surveillance does have some merit when the stream of data being analyzed includes the results of real-time water-quality monitoring. When these results are used as the primary means of detecting an attack and the other subsidiary data are used as confirmatory and supporting, the approach has considerable merit, but the idea of waiting for casualty counts before remedial action is taken in the event of a chemical attack is irresponsible and unacceptable.

The Value of Monitoring

Monitoring is a critical component of any water security program. There is no other feasible way to address the severe vulnerability presented by the threat of an intentional contamination event, especially in the distribution system. The early detection of any such event is imperative, to decrease the horrendous potential for mass casualties. Nevertheless, while preservation of human life is the number one priority, it is not the only imperative.

Containment and isolation is critical in limiting casualties, but it is also imperative to limit the cleanup of any incident. The anthrax cleanup for the Hart Senate Office Building after the contaminated mail incident cost the EPA over $27 million from its Superfund budget.[18] It is possible that some agents will not be able to be cleaned up, and piping will need to be replaced. This could be a very expensive proposition, considering that not only the main pipes but also some household plumbing may need to be replaced. Further, if the agent is widely disseminated within buildings, owing to aerosolization, entire structures may need to be abandoned. Therefore, rapid detection and containment are critical in reducing casualties and in limiting cleanup costs.

With the current state of technology, there is no need for us to operate our water systems as if blindfolded. Admittedly, the instrumentation available today isn't going to give us x-ray vision (or perhaps even 20/20 vision), but it will allow us a clear enough picture to avoid many of the hazards that we would surely encounter if we left the blindfold securely in place.

Notes

1. Buckley, Chris. 2005. Toxic water surge threatens millions. *Herald Sun*. November 25.
2. Kroll, Dan. 2005. Utilization of a new toxicity testing system as a drinking water surveillance tool. In *Water Quality in the Distribution System*. Denver: American Water Works Association.
3. Speiser, O. H., W. Scholz, G. Staaks, and D. Bagnaz. 1996. The influence of cyanotoxins on the behavior of zebrafish (*Brachydanio rerio*): Methods and results. Presented at Measuring Behavior '96—International Workshop on Methods and Techniques in Behavioral Research. Utrecht, The Netherlands. http://www.noldus.com/events/mb96/abstract/spieser.htm

4. Aquatic Biomonitoring. USACHER. http://usacehr.detrick.army.mil/Aquatic%20Bio monitor%20Product.html

5. Moldaenke, C. F. Real time biomonitoring to check water quality. National Water Monitoring Conference. Madison, WI. May 19–23, 2002. http://water.usgs.gov/wicp/acwi/monitoring

6. bbe Moldanke. http://www.bbe-moldaenke.com/EN/Biomonitoring/Daphnia_Fish_Toximeters.html (accessed December 8, 2005).

7. McCarter, Steven M., and Adonis C. Cassinos. 2005. Utilization of naturally occurring biosensors to detect toxins in our nation's water supply. United Defense, December 9. http://www.uniteddefense.com/WaterSentry_WhitePaper.pdf

8. CheckLight. http://www.checklight.co.il/ (accessed December 9, 2005).

9. s::can. http://www.s-can.at/index.php?id=12 (accessed December 14, 2005).

10. Sandia National Lab. http://www.ca.sandia.gov/chembio/microfluidics/detection/ (accessed December 15, 2005).

11. JMAR. http://www.jmar.com/2004/about.shtml (accessed December 21, 2005).

12. EPA ETV Program Run by Battelle. 2004. Test QA plan for verifying multi-parameter water monitors for the distribution system. http://www.epa.gov/etvprgrm/pdfs/testplan/01_tp_monitors.pdf#search='haught%20roy%20water%20response'

13. Byer, David, and Kenneth H. Carlson. 2005. Real-time detection of intentional chemical contamination in the distribution system. *Journal of the American Water Works Association.* 97 (1): 58–61.

14. Kroll, Dan. 2002. Results of threshold beaker testing on chemical treat agents: Is on-line water security monitoring feasible?" Internal Hach HST document. September 12.

15. EPA. 2006. Environmental Test Verification (ETV). http://www.epa.gov/etv/verifications/verification-index.html

16. Englehardt, Terry. 2005. E-mail message to author. November 7.

17. EPA. 2006. Threat Ensemble Vulnerability Assessment (TEVA) Computational Framework Fact Sheet. http://www.epa.gov/nhsrc/pubs/fsTEVA111505.pdf#search='threat%20ensemble%20vulnerability%20assessment%20fact%20sheet'

18. Ramstack, T. 2003. GAO scores contractor work, pay in Hart anthrax clean-up. *Washington Post.* June 18.

8

Responding to an Event

In preparing for battle I have always found that plans are useless, but planning is indispensable.

—Dwight D. Eisenhower

Let our advance worrying become advance thinking and planning.

—Winston Churchill

The Dilemma

The problem with detecting a water contamination emergency, whether terror related or accidental, is how to respond effectively to limit the toll on life and property. The simplest answer is that if we suspect a problem, we will simply shut off the water. This is an unacceptable answer for many reasons. It is unlikely that the public would be accepting of frequent disruptions in their water supplies for false alarms. Moreover, if the morning shower of the mayor of a community were interrupted without good reason, the utility manager of that district would most likely be looking for a new job. Thus, certainty is required before taking any action as drastic as stopping supplies.

Some utilities do not have the option of shutting down the supply. Some water delivery pipes are in such a poor state of repair that the reduced pressure that would result from disruption of the water supply could lead to major pipe failure. Other utilities, especially in large metropolitan areas, cannot shut down because of the necessity of maintaining a pressure head for fire suppression and other critical functions. However, if no action is taken, there is the grave risk that an event will cause mass casualties. The problem then is basically one of controlled response. We must carefully weigh and balance the problems of over-responding against those of under-responding, to ensure that proper steps are taken during each phase of a potential emergency.

EPA Guidance

The EPA has performed extensive research in the area of response protocols and has developed a number of useful tools for formulating response plans. Chief among these is a manual, available on the Web, entitled "Response Protocol Toolbox: Planning for and Responding to Drinking Water Contamination Threats and Incidents,"[1] subdivided into six modules (see table 8–1).

Table 8–1. EPA toolbox contents

Module	Contents
1	Water Utility Planning Guide
2	Threat Management Guide
3	Site Characterization and Sampling Guide
4	Analytical Guide
5	Public Health Response Guide
6	Remediation and Recovery Guide

The protocol recommends action be initiated on the basis of the incident command system (ICS). Under the ICS, the utility names a water utility emergency response manager (WUERM), who takes control of an incident from its onset. The EPA manual provides very detailed guidance and includes copies of many forms and procedures for use in initiation of a response. A brief synopsis of the EPA protocols is presented here.

Possible threats

To ensure that proper actions are taken and that analysis is performed throughout the course of an incident, the guidance advocates a system that utilizes three main threat levels. The first and lowest of these threat levels is deemed a possible threat. This is when a determination is made regarding whether a threat is possible. Indications of a possible threat to the water supply can come from a variety of different sources—including a physical security breach, witness account, notification by the perpetrator, notification by law enforcement, notification by the news media, consumer complaint, public health information, or unusual water-quality readings (e.g., see fig. 8–1).[2]

In this discussion, unusual water-quality data is the primary source of the initial report of a possible incident. In this case, there is a possible threat when there is a significant deviation in water quality from the normal or baseline for a given location. This indication of unusual quality could be provided through any of the online monitoring techniques described in chapter 7. It could also be provided through routine sampling or based on customer complaints.

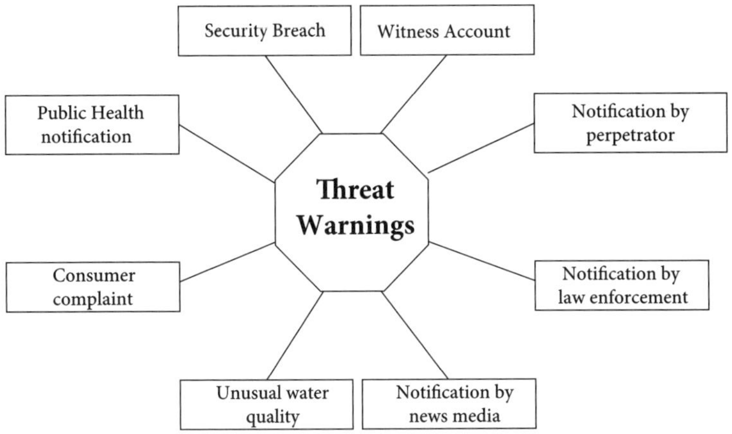

Fig. 8–1. Indications as to a possible threat

Once a possible threat has been detected, the first step is to make sure that the unusual water-quality readings are not the result of normal, day-to-day operational procedures. Multiparameter water-quality meters, equipped with algorithms that can learn and store signatures of normal water-quality events, are a great advantage in making these determinations. As previously discussed with regard to institutional intuition, this type of information can actually be utilized in streamlining and improving operational procedures, resulting in reduced variation in water quality over time.

For example, consider a real incident that occurred in a major East Coast city just downstream from a water storage tank. A multiparameter water panel distribution monitor and an event monitor algorithm were installed at this location, and the event monitor recorded a regular alarm (fig. 8–2). Careful evaluation of the baseline parameter data showed that a chlorine upset was triggering the alarms (fig. 8–3). The chlorine levels would gradually rise over time and then suddenly drop; it was this sudden drop that was triggering the alarm.

Further investigation revealed the cause of these upsets. The storage tank was normally filled from water source A, but at times of peak demand, both source A and source B were turned on to fill the tanks. Source B has a higher chlorine residual than source A, so when it was used to fill the tank, the chlorine slowly increased. When source B was turned off, owing to hydrodynamic short-circuiting, the chlorine level decreased rapidly to match the concentration of source A. After this was determined to be the cause, the learning capability of the algorithm was used to name and classify this event as benign, so that when it occurred again, the alarm was recognized. The screen would report the event as "Name: Pump Shut Off, Type: NORMAL." Information such as this could be utilized to reduce the occurrence of such events.

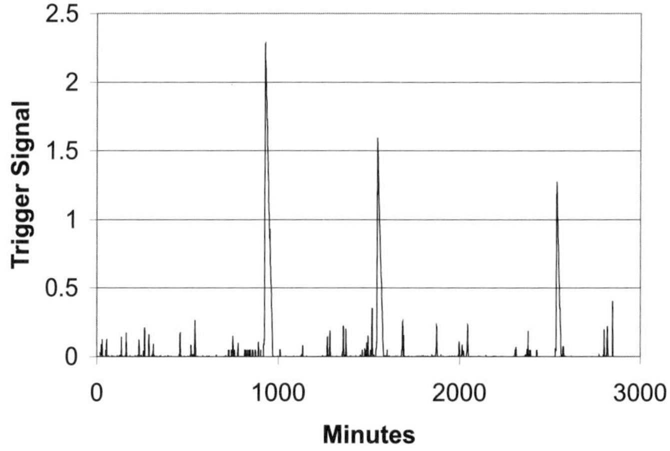

Fig. 8–2. East Coast location showing regular alarm events

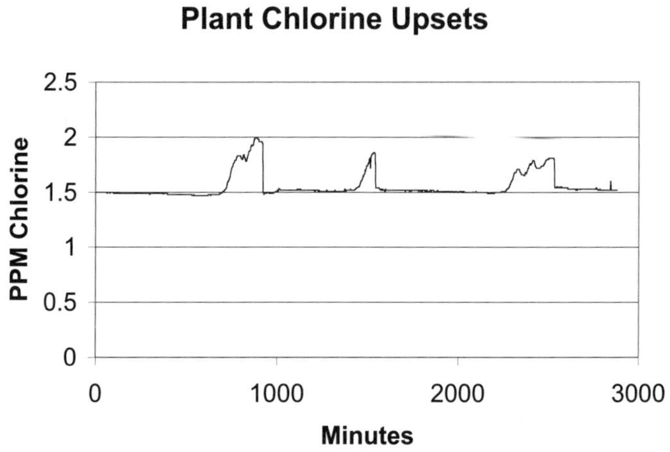

Fig. 8–3. Chlorine upsets triggered the alarms in figure 8–2.

As another example, in a plant that uses caustic feed to control water pH, a trigger alarm event was noted. Investigations showed that the alarm was the result of a caustic overfeed. The root cause of the overfeed was that the vendor had delivered the wrong concentration of caustic to the plant. Detection of this event through monitoring allowed for the instigation of controls for incoming chemicals to prevent a recurrence (fig. 8–4).

If no normal or operational causes of changes in water quality are indicated, other possible causes should be investigated. For example, are other monitoring sites responding in a similar manner, and is the response distribution such that it could indicate a change in the source water? Many systems rely on multiple source waters for their supplies, and switching from one source to another can dramatically change water-quality readings. This is another case in which events could be interpreted by a heuristic learning-based system.

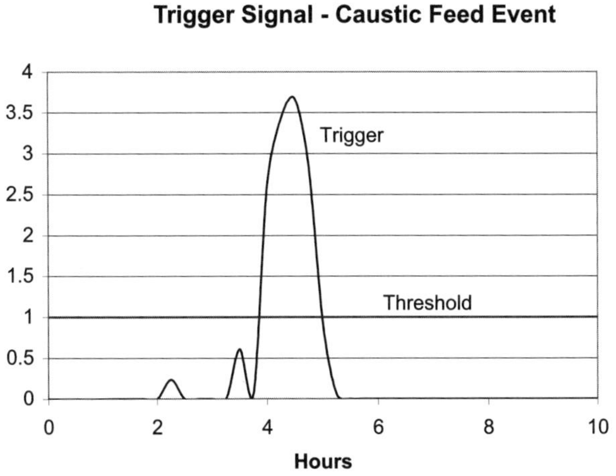

Fig. 8–4. Caustic overfeed event detected through online monitoring

If the cause of the readings is not apparent, further investigation is warranted through initiation of a site-characterization study. For this purpose, the water is rapidly tested on site at the chosen locations or samples are collected for more detailed laboratory analysis.

Transition from possible to credible—site characterization

When anomalous water-quality readings are observed in conjunction with other information such as customer complaints, threats, and reported illnesses, the key to the transition from possible to credible threat, the next stage in the emergency hierarchy, is site characterization. Module 3 of the EPA protocol offers guidance on the key testing to confirm the possibility of a significant water-quality event.[3]

The EPA recommends that testing be broken into two tiers: core field testing and advanced field testing. The EPA recommendations for core field testing address radioactivity, cyanide, chlorine residual, pH, and conductivity (fig. 8–5). These are common parameters, and quick, easy-to-use field methods are available for the onsite testing of all the listed parameters:

- The radiation test is basically a site surveillance test, to ensure that it is safe for the personnel who are testing the water to enter the area.

- Cyanide is a very common threat agent. Testing for cyanide is easily accomplished by either ion-selective electrodes or colorimetric methods. (Note, however, that if the chlorine residual test is run first and found to be normal, then there is no need to run the cyanide test, because chlorine rapidly removes active cyanide complexes.)
- The chlorine residual test may indicate contamination with any compound that reacts with chlorine, including organics and bacteria (however, not all organics have a significant chlorine demand).
- pH and conductivity can be indicators of contamination with inorganics. If a multiparameter online monitoring device provides the initial indication of trouble, it would be advisable to include field methods, in addition to the core kit recommendations from the EPA, to verify that the parameters being monitored online are reading correctly.

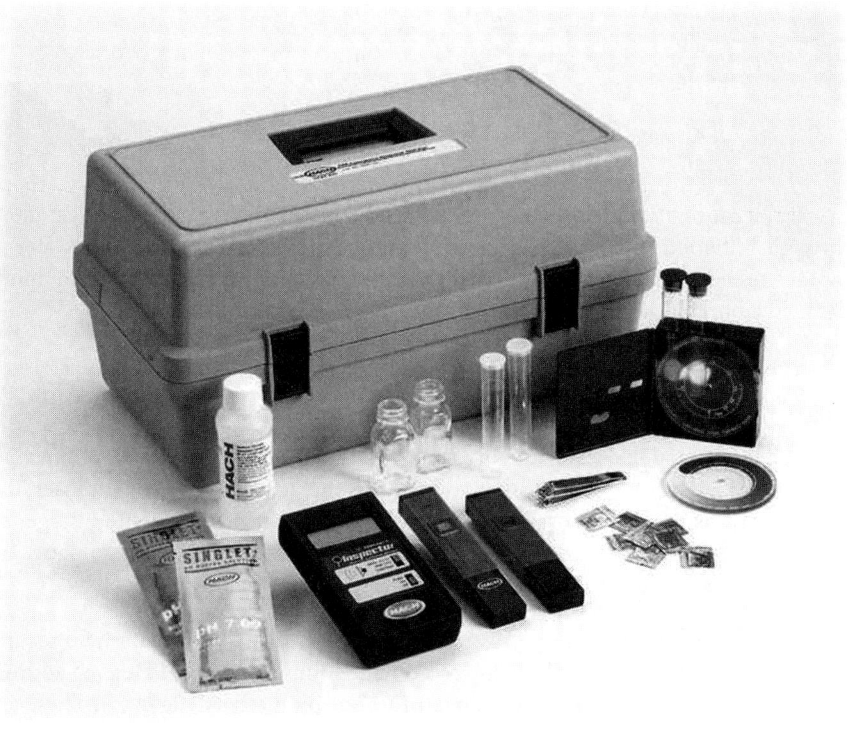

Fig. 8–5. EPA Core Emergency Response Toolkit, with tests for radiation, cyanide, chlorine residual, pH and conductivity

Depending on the situation, a responder may wish to perform more advanced field testing on a given site. Various tools, kits, and instruments are readily adaptable to expanded field testing. The EPA recommends several tests that may be performed as part of expanded field testing (table 8–2).[3]

Table 8–2. Expanded field test kits

Parameter	Class	Method	Comments
General hazards	Safety screen	HazCat	Trained HazMat responder
Volatile Chemicals	Safety Screen	Sniffer-type devices	Detects chemicals in air
Chemical Weapons	Both	Enzymatic/ colorimetric	May also detect certain pesticides
Water quality parameters	Water	Various	Many kits for common parameters
Pesticides (OP and carbamates)	Water	Immunoassays	Quick and easy to use
VOCs and SVOCs	Water	Portable GC/MS	Expensive
Biotoxins	Water	Immunoassay	Quick simple
Pathogens	Water	Immunoassay and PCR	Pre-concentration may be needed
Toxicity	Water	Inhibition of biological activity	Need to know baseline reading for site

Many of these expanded methods are new and unproven; thus, caution should be the rule until confidence as to their ability to provide accurate results has been established. The EPA runs a program known as Environmental Technology Verification (ETV), which independently evaluates such technologies. Evaluation by the ETV program doesn't certify or endorse the verified products, but provides an independent source of data related to performance under the specified test conditions. A description of the ETV program and the technologies that have been verified can be found on the ETV Web site.[4] Brief descriptions of technologies used in site evaluation, some of which have been evaluated through the ETV program, are given in the next sections.

Toxicity tests

A variety of methods are available for field toxicity testing, relying on a number of different organisms and methodologies to determine the relative toxicity of a water sample. A notable drawback of toxicity testing is that it provides only a relative measure, which can vary considerably by location. Knowledge of a specific location's baseline toxicity is imperative when looking for significant deviations.

Importantly, the toxicity of a water source that is used as a blank can vary. This includes deionized water, since in the course of the deionization process, the resins can begin to release materials that cause a change in toxicity. Therefore, an apparent change in a sample's toxicity may be due to a change in the blank standard's toxicity. It is a good idea to maintain a large supply of the baseline water against which all future testing is performed.

Chemiluminescence. One method of monitoring toxicity is based on chemiluminescence. This method is used in the Severn Trent Services Eclox kit (fig. 8–6). The reaction of luminal and an oxidant, in the presence of horseradish peroxidase, results in the chemical production of light or chemiluminescence, which can be used to detect the presence of toxins. Any free radical scavengers or antioxidants, such as those contained in feces or urine, will interfere with the reaction, reducing the light output. Substances such as phenols, amines, heavy metals, or compounds that attack or coat the enzyme will also reduce the light output. The light output is plotted over time and produces characteristic curves. Results are compared to deionized water. Samples containing pollution give lower light levels.

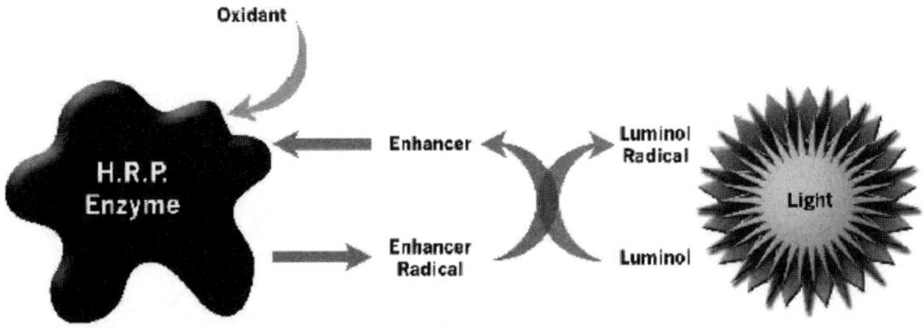

Fig. 8–6. The Severn Trent Services Eclox kit detects toxicity on the basis of inhibition of the enzyme horseradish peroxidase, thereby impeding the production of light from the chemical luminal. (Drawing courtesy of Severn Trent Services)

The Eclox method is extremely robust and has been hardened for field use by the British military (fig. 8–7). This method has been shown to be effective against a wide variety of substances that could be potential risks (e.g., see table 8–3). The kit is commercially available, packaged with subsidiary tests for pH, conductivity, color, arsenic, and cholinesterase-inhibiting substances, including pesticides and nerve agents. The cholinesterase-inhibitor test is described later in this chapter. (The effect of chlorine can be counteracted with a pre-conditioner contained in the kit.)

Fig. 8–7. The Eclox kit is provided with a variety of subsidiary tests designed by the Hach Company to supplement the toxicity measurement. It has been hardened to military specifications.

Table 8–3. Response of Eclox to substances with different modes of toxic action (Source: *Internal report, Wolfson Applied Technology Laboratory and Severn Trent Services)*

Mode of Action	Compound	Limit of Detection (mg/L)
Polar narcotics	Phenol	0.06
	Aniline	0.04
Respiratory blockers	Cyanide	0.005
Oxidative uncouplers	2,4-dinitrophenol	0.7
Membrane irritants	Chlorine*	0.004
	Acrolein	0.9
Cholinesterase inhibitors	Carbofuran	0.5
CNS convulsants	Endosulphan	1.2
Heavy metals	Arsenic	38
	Copper	1.0
	Mercury	1.0
	Chromium	92
	Lead	5.0
	Thallium	5.0
	Antimony	5.0
		7.0
		7.0
Polysynthetic inhibitors	Prometon	900
	Bromacil	210
Cell division inhibitors	Triflualin	4

Bioluminescence. Another way to monitor toxicity is based on the production of light by various species of luminous bacteria. CheckLight of Israel has developed a system based on the inhibition of light production by the bacteria *Photobacterium leiognathi.* This system has the ability to use different buffer screening systems to differentiate between organic pollutants and metal toxins.

More commonly, the luminescent bacteria *Vibrio fischeri* is used. This bacteria is the basis of the BioTox Flash Test produced by Hidex Oy of Finland and the MicroTox and DeltaTox (fig. 8–8) test kits produced by Strategic Diagnostics. These kits have been shown to respond to a wide variety of different compounds. They all utilize the reduction of light output as measured by a luminometer, a fairly expensive piece of equipment, to detect toxicity to the bacterial cultures. Problems with removal of chlorine in finished drinking water samples, though, may skew results.

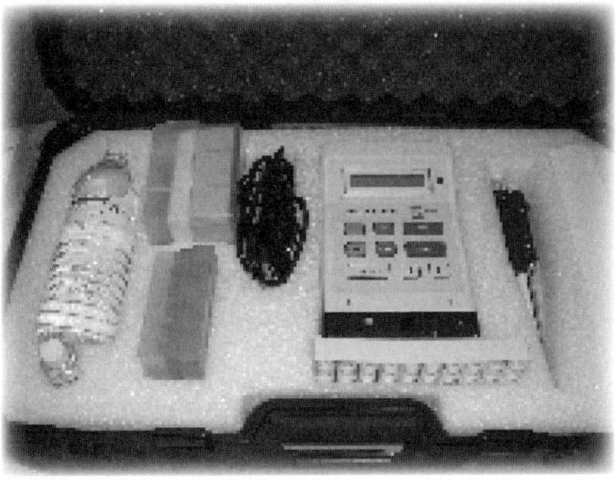

Fig. 8–8. DeltaTox uses the inhibition of light produced by the luminescent bacteria Vibrio fischeri *to indicate toxicity.*

Bacterial respiration. Another method for measuring toxicity is the inhibition of bacterial respiration. PolyTox, from InterLab Supply, uses standard dissolved oxygen electrodes to measure respiration of a specially formulated bacterial culture. Oxygen consumption is a good measurement of overall bacterial health; however, this method is not without problems: the test requires a dissolved oxygen electrode, which is a touchy piece of equipment; the samples must be aerated for 30 minutes before staring the test, which can result in the loss of volatile toxins from the sample; and only one test can be run at a time (multiple tests require long periods of time coupled with extensive sample manipulation).

Another method based on bacterial respiration is the ToxTrak Rapid Toxicity testing system, produced by the Hach Company. Rather than directly measuring oxygen consumption, the ToxTrak system utilizes a colorimetric system based on the rate of reduction of the dye resazurin (fig. 8–9). As the bacteria actively metabolize, the dye is reduced from blue to pink.

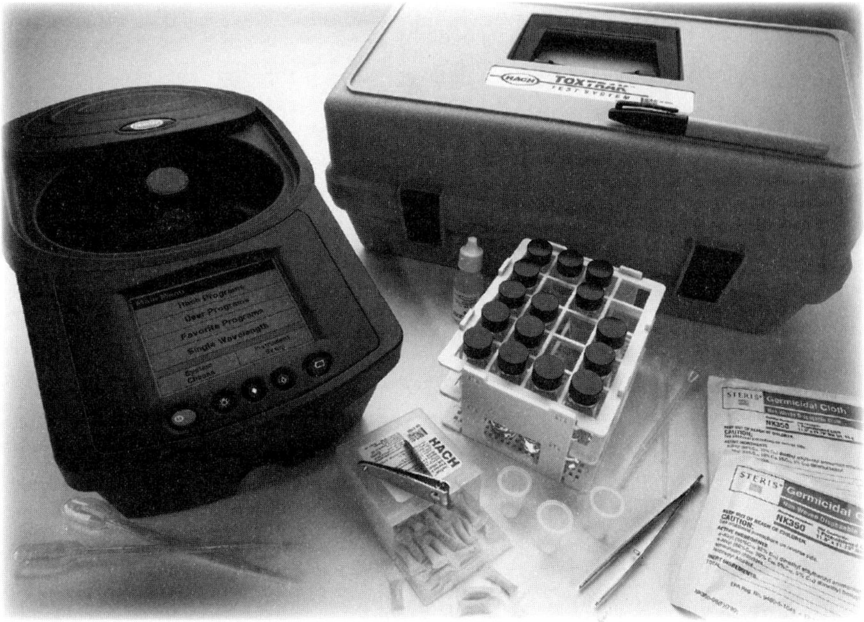

Fig. 8–9. The Hach ToxTrak system can utilize any bacteria culture with any spectrophotometer to detect toxicity.

Bacterial respiration is normally a slow process, but the Hach ToxTrak system makes use of a patented accelerator solution, increasing the rate of reaction and allowing the test to be completed in as little as 45 minutes. Inhibition of the rate of reduction indicates the presence of toxicity. The color change can be measured with any spectrophotometer or colorimeter capable of measuring at 600–610 nm (fig. 8–9). The color change can also be measured visually, and a color disk comparator method is available. This system is inexpensive and has been shown to react to a wide variety of toxins (e.g., see table 8–4). Although in theory virtually any bacteria may be used in the test, cultures must be grown and maintained in advance of testing.

Table 8–4. Reaction of bacterial inhibition methods to various compounds

Toxin	PolyTox™ % Inhibition	ToxTrak™
Cu^{2+} 2 mg/L	31%, 18%, 56%	52%, 61%, 59%, 57%, 68%, 57%, 73%, 63%, 73%
Cu^{2+} 0.2 mg/L	–4%	53%, 48%, 56%, 49%, 73%, 55%, 59%, 58%, 64%
Cu^{2+} 0.02 mg/L	4%	–7%, 8%, –6%, 3%, 0%, –12%
Formaldehyde 0.5%	89%, 87%	87%, 86%, 92%
Formaldehyde 0.05%	50%, 37%, 0%, 55%	10%, 28%, 24%, 23%
Formaldehyde 0.005%	8%, 25%	9%, 13%
o-Chlorophenol 30 mg/L	0%	0%
Centralia, WA. Wastewater	–8%, –23%	15%, 17%
Hg^{2+} 10mg/L	68%, 50%	47%, 55%, 42%
Hg^{2+} 5 mg/L	0%, 0%	22%, 36%, 38%
Hg^{2+} 1 mg/L	–12%	–4%, 15%, 4%
Grotan HD2 1.0%	90%	67%, 82%, 40%, 65%
Grotan HD2 0.1%	50%	20%, 16%
Grotan HD2 0.01%	0%	10%, 16%, –7%, 14%, –13%, 8%
Phenol 100 mg/L	–16%	40%
Phenol 10mg/L	–8%	1%
CN^- 10 mg/L	36%	28%
CN^- 1.0 mg/L	18%	22%
CN^- 0.1 mg/L	9%	10%
10 mg/L phenol + 2.0 mg/L Cu^{2+}	0%	60%
1 mg/L phenol + 0.2 mg/L Cu^{2-}	0%	57%
0.1 mg/L phenol + 0.02 mg/L Cu^{2+}	0%	0%

Testing with *Daphnia* and other invertebrates. Invertebrates such as the water flea *Daphnia* have long been a mainstay of the toxicology laboratory. Strategic Diagnostics offers a variety of test kits that use a variety of invertebrates, including *Tetrahymena thermophila, Brachionus calyciflorus, Brachionus plicatilis, Thamnocephalus platyurus,* and several *Daphnia* species. Toxicity in these tests is based on the success or failure to ingest red microspheres that are clearly visible when present in the organism's digestive tract. This is a much easier end point to judge than are found in traditional assays that look for lethality or changes in behavior.

Aqua Survey of Flemington, New Jersey, has simplified classic invertebrate toxicity testing in their IQ Toxicity Test. This method is based on fluorescent tagging of a sugar molecule that is placed in the *Daphnia*'s food. If the *Daphnia* are happy and actively metabolizing, they will ingest the sugar and cleave the molecule. This cleaving of the molecule causes the organisms to become fluorescent under UV light. It is basically a variation of the classic 4-methylumbelliferyl-β-D-glucuronide (MUG) test for the bacteria *Escherichia coli.* If the *Daphnia* glow, then there is no toxicity; if they don't glow, then there is toxicity. *Daphnia* testing is extremely sensitive and may even be overly sensitive to certain common drinking water constituents.

The largest drawback of all forms of invertebrate testing is culture maintenance. For emergency testing, a usable culture needs to be maintained at all times. In some tests, organisms need to be in a specific stage of development or state of hunger. This is hard to maintain for an emergency program and is probably better suited to ongoing laboratory testing, in which the organisms can more easily be maintained in the proper state.

All methods of toxicity testing require knowledge of a baseline. Also, some toxicity-based methods may be too sensitive to employ in the distribution system. Water treatment chemicals and common constituents of drinking water, such as trace metals that are not toxic to humans, may adversely affect invertebrates.

The choice of whether to use toxicity methods depends on the resources available. A dedicated program that maintains some knowledge of baseline conditions can take a sizeable amount of resources. Also, because of the differences in response to various toxins elicited by different toxicity methods (see table 8–5), it is a good idea to utilize more than one method or back them up with subsidiary tests, such as those provided in the Severn Trent Eclox kit.

Table 8–5. Results of ETV evaluation of several types of toxicity test

			Manufacturer							
			Severn Trent Services	Strategic Diagnostics	Strategic Diagnostics	Hach	Inter Lab Supply	Hidex OY	Aqua Survey	Check Light
System		Lethal Dose mg/L	Eclox	Deltatox	Microtox	ToxTrak	POLYTOX	BioTox	IQ Toxicity Test	ToxScreen II
Contaminants in DDW	Aldicarb	280	280	28	28	280	ND	ND	3.5	0.28
	Colchicine	240	24	ND	240	240	ND	ND	25	0.24
	Cyanide	250	0.25	0.25	0.25	25	0.25	25	0.25	0.25
	Dicrotophos	1400	1400	140	140	14	ND	ND	0.88	0.14
	Thallium Sulfate	2400	2400	240	240	2.4	2400	24	120	ND
	Botulinum Toxin	0.3	ND	ND	ND	ND	ND	ND	0.0003	0.03
	Ricin	15	15	ND	ND	0.015	15	ND	0.015	15
	Somam	0.15	ND*	ND	ND	ND	ND	ND	0.0013	ND
	VX	0.22	0.49	ND	ND	0.22	ND	ND	0.0095	ND
			Inhibition Values							
Potential Interferences	Aluminum 0.36 mg/L		–2	3	1	–3	–8	16	90	16
	Copper 0.65 mg/L		4	38	61	-6	5	96	100	59
	Iron 0.069 mg/L		2	–3	–5	36	7	0	90	29
	Manganese 0.26 mg/L		62	–2	9	11	6	10	3	–23
	Zinc 3.5 mg/L		10	22	28	–17	11	48	7	–68

Immunoassays

Bacteria and biotoxins are not always detectable through conventional means or toxicity testing. Immunoassays exploit the interaction between antibodies and antigens to detect the presence of specific organisms or compounds. These tests come in a variety of formats and are capable of detecting a wide variety of bacteria, viruses, biotoxins, and specific chemicals.

One of the most common immunoassays is a lateral flow assay (fig. 8–10). We are all familiar with this format, as it is the same one that is used in commercially available home pregnancy tests. After a sample is applied to the reservoir end of the strip, the liquid begins to migrate down the length of the strip. In the course of moving down the strip, it comes into contact with regions of the test strip that are impregnated with specific antibodies for the antigen being tested. These antibodies are tagged with colored or fluorescent labels. They reach a certain area in the strip that binds and captures them at that site using other or the same antibodies. The result is a colored or fluorescent line if the target antigen is present. A control line is usually included, to verify that everything has run properly. These tests are quite simple to use and can usually be read with the naked eye.

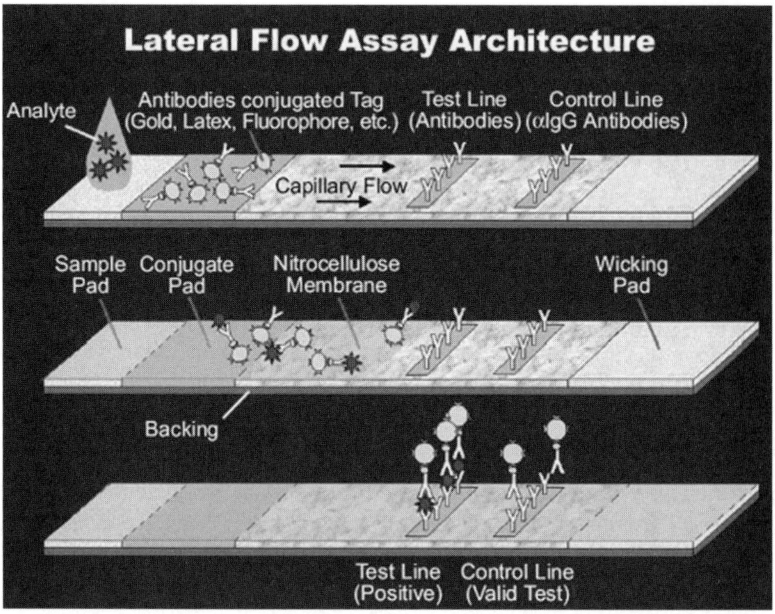

Fig. 8–10. The quantitative lateral flow assay (QLFA) is a test strip for identifying biological organisms in an analyte (liquid sample), which flows from the sample pad to the wicking pad. On the conjugate pad, specific antibodies (Y shapes) tagged with chemical markers (ovals) bind to the target antigen (sunbursts) in the sample and flow toward the wicking pad. At the test line, other immobilized antibodies bind the antigens to produce a positive test result, which is revealed as fluorescing color. A control line antibody confirms that the test ran successfully (i.e., the sample flowed the length of the test strip). (Courtesy of NASA)

A number of manufacturers produce these test strip assays for analytes of interest. One problem with this sort of assay is that they are very specific for the antigen being tested. Therefore, you need to know what you are looking for and run the appropriate test. Another problem is specificity. Cross-reactivity with other microbes or antigens can result in false positives.

A further problem with these tests is that their sensitivity is lower than could be desired. Certain compounds, such as botulinum toxin, can be detected at fairly low levels, while others, such as ricin, require more of the compound to be present; also, as shown in table 8–6, bacterial detection levels are usually fairly high. Thus, it is advisable to use a sample pre-concentration device or pre-culture before performing the test. Several pre-concentration methods are currently under study but they are currently commercially unavailable. Because pre-culture methods can be time consuming, they may not be the best choice for emergency response situations.

Table 8–6. Levels of detection of various antigens by use of Bio Threat Alert Strips, produced by Tetracore of Gaithersburg, MD (Source: www.tetracore.com/products/domestic.html)

Antigen	Limit of detection
Bacillus anthracis	1 x 10^5 cfu/mL
Yersinia pestis	2 x 10^5 cfu/mL
Francisella tularensis	1.4 x 10^5 cfu/mL
Botulinum toxin	10 ppb
Staphylococcal enterotoxin B	2.5 ppb
Ricin	50 ppm

cfu = colony forming unit

Test strips for pesticides and nerve agents

Another type of test strip can be used to detect nerve agents and pesticides. Enzyme-based test strips are available from Severn Trent Services; these test strips work on the basis of inhibition of the enzyme cholinesterase to change the color of a dye (fig. 8–11). Many nerve agents and pesticides are capable of inhibiting this change and are thus detected by the strips.

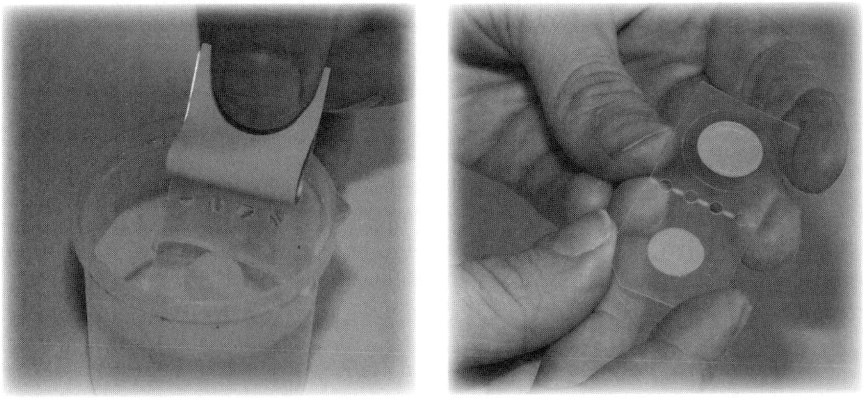

Fig. 8–11. Cholinesterase test strips are simple and easy to use. (Courtesy of Severn Trent Services and the Hach Company)

The test is a qualitative test for the detection of pesticides on the basis of cholinesterase inhibition (see table 8–7). One side of the ticket contains a disk that is saturated with cholinesterase, an enzyme present in most living organisms, except plants, the main function of which is to control muscle performance. If the enzyme is altered or dies, so does the organism. Insecticides can inhibit an organism's ability to produce cholinesterase and therefore kill the organism. If enough insecticide is present in the tested sample, it will inactivate the cholinesterase that is chemically bonded to the ticket and prevent a chemical reaction, which when insecticides are absent turns the disk blue. White color indicates the presence of insecticides. In the absence of pesticides, the cholinesterase hydrolyzes an ester to form a colored compound. Pesticide inhibition will interfere with the reaction, stopping hydrolyzation and color development on the test strip.

Table 8–7. Typical pesticide test strip detection limits (Source: Data provided by Severn Trent Services)

Compound	Detection Limit in Water (mg/L)
Carbamates	
Aldicarb	0.2
Carbaryl	7.0
Carbofuran	0.1
Mesurol	5.0
Methomyl	1.0
MIPC	2.0
Oxamyl	1.0
Propoxur	1.0
Orgnaophosphates	
DDVP	3.0
Methamidophos	4.0
Mevinphos	2.0
Thophosphates	
Asphon	5.0
Azinphos-methyl	0.3
Chlorpiyrifos-Ethyl	0.7
Chlorpyriphos-Methyl	1.0
Diazinon	2.0
EPN	0.2
Fenitrothion	1.5
Malathion	2.0
Metasystox-R	20.0
Methyl-Parathion	4.0
Parathion	2.0
Phorate	3.0
Phosmet	1.0
Phosvel	0.8

Gas chromatography

For a detailed description of GC, see chapter 7. Manufacturers have increasingly been deploying GC in portable field-based systems, for security checks. These systems usually rely on a purge-and-trap technique, in which the volatile compounds are purged from the water with an inert gas and are then trapped on an organic resin from which they are desorbed into the instrument. GC is part of a wide variety of hyphenated techniques (table 8–8) that can improve the sensitivity of the GC to various analytes.

Table 8–8. Hyphenated GC techniques

GC-TCD	Thermal conductivity detector
GC-SAW	Surface acoustic wave
GC-ECD	Electrolytic conductivity detector
GC-FID	Flame ionization detector
GC-PID	Photo ionization detector
GC-MS	Mass spectrometer

One version of portable GC is HAPSITE, produced by Inficon, of East Syracuse, New York. This instrument enjoys widespread use in the military and environmental fields. It utilizes a purge-and-trap technique in a special sample attachment that allows the analysis of water samples in the field. Although it is an expensive piece of equipment that is limited to the detection of volatile organics, HAPSITE can detect a wide variety of these volatile compounds and thus can be a useful tool for field testing.

Infrared spectroscopy

The limitations of GC (which detects volatile organics only) may be overcome by using GC in conjunction with infrared (IR) spectroscopy. The use of IR spectroscopy has traditionally been confounded by direct interference of the water in the samples with the test method; consequently, aqueous samples that contained less than about 10% product were not measurable by IR spectroscopy.

To expand the capabilities of IR spectroscopy, Smith Detection's SensIR Technologies initially produced a portable IR spectrophotometer, called the HazMat ID, that uses Fourier transform IR attenuated total reflection spectroscopy. HazMat ID is designed to detect and identify WMD, TICs, narcotics, and explosives in nonaqueous samples. A system called the ExtractIR was developed to extend the capabilities of the HazMat ID to aqueous samples. ExtractIR allows the extraction of nonvolatile organic compounds from the water, so that they can be analyzed via IR spectroscopy. Even with this extraction system, detection limits are fairly high, on the order of 100 ppm in water.[5]

Detection of adenosine triphosphate

Adenosine triphosphate (ATP) is the component of cells that is responsible for energy transfer. Therefore, all living cells contain ATP. Since this is the case, changes in ATP levels in water should be usable as an indication of biological contamination. ATP measurement has long been used in the clean room industry, as an indicator of proper cleaning and sterilization of work surfaces.

Several manufacturers have adapted this test for use in water samples. These systems are usually based on the role that ATP plays in providing the energy source for bioluminescence. This is a good candidate method for detecting gross changes in the level of ATP present in a sample.

The problem is that the water in our distribution system is not sterile and always contains some level of ATP. Thus, it is imperative to know the baseline, as in toxicity testing. There is also the problem of distinguishing whether a rise in ATP levels is due to a general sloughing off of biofilm or to a bacterial attack. Differential methods for lysis of the cells of certain organisms are currently under study, to render this method more specific for designated types of bacteria.

Polymerase chain reaction

Polymerase chain reaction (PCR) is a technique for detecting living organisms by extracting and multiplying the DNA specific to that organism. The technique allows a small amount of the DNA molecule to be copied, amplifying it exponentially. Once more DNA is available, analysis becomes much easier. The DNA can be detected via a variety of methods, including fluorescent gene probes, fluorescent melting curves, or electrophoresis. PCR is commonly used in medical and biological research laboratories for a variety of tasks. PCR as it was originally devised is not a simple procedure and normally relies on advanced laboratory techniques.

In recent years, manufacturers have developed a variety of automated techniques that decrease the amount of expertise required in order to perform this technique, allowing for use by relatively unskilled technicians in the field. One automated PCR methods is the Ruggedized Advanced Pathogen Identification Device (RAPID), from Idaho Technologies of Salt Lake City. RAPID uses freeze-dried reagents and can simultaneously screen for up to eight different targets, chosen from a list of potential viral or bacteria pathogens. This very sensitive technique can detect the presence of even a small number of organisms in a sample (e.g., as few as five *Bacillus anthracis* cells).

An even smaller automated PCR device produced by Idaho Technologies is the Razor. Razor can detect up to 12 analytes at once, also using freeze-dried reagents. Both of these instruments are rugged and easy to use and are capable of detecting very low levels of the analytes in question. The main problem, however, is price. A RAPID unit costs around $55,000, with reagents for each test running $50.

Multiparameter lab-on-a-chip technologies

Multiparameter monitors reduced to a lab-on-a-chip configuration are becoming available for field testing. One is the WaterPOINT 855 multiparameter water quality analyzer, produced by Sensicore of Ann Arbor, Michigan. This chip-based instrument is a handheld device capable of measuring 14 separate parameters in one quick test, including pH, ORP, conductivity, total dissolved solids, temperature, free chlorine, total chlorine, monochloramine, calcium, total hardness, carbon dioxide, total alkalinity, ammonium, and the Langelier saturation index.

Research is underway to develop a communication module that directly downloads field measurements to a central location where they can be compiled in a real-time decision-making mode. This instrument offers the capability of quickly measuring several parameters at once. Drawbacks are that many of these parameters are not independent—for example, conductivity and total dissolved solids are directly related. Also, many of these parameters have little or no bearing on security monitoring; performing superfluous tests adds little or no value to an instrument and only increases the cost per test. It may be simpler and more cost-effective to separately perform conventional tests, using simple electrochemical or colorimetric methods.

Technologies on the horizon

Massive development and research is currently focused on field methods for detecting various threat agents. Much of this work has been on the air side of the threat matrix, with little emphasis on water. As water has gained credibility as a potential threat, research into adapting many of the existing methods and developing new methods to handle aqueous samples has begun. These methods include ion mobility spectroscopy (IMS), surface acoustic wave (SAW), Raman spectroscopy, quantum dot technology, and micro arrays, and other new technologies. As these technologies are developed and tested, the available number and capability of tools for determining the presence or absence of harmful materials in our water will increase.

Credible threats

According to the EPA, a water contamination threat is considered to be credible if additional information collected during the investigation corroborates the threat warning and if the collected information as a whole indicates that contamination is likely. Additional indications such as strange water-quality readings, without an obvious operational cause coming from site studies conducted after an initial reading from online instruments, may be sufficient to place the incident into the credible range.

Time is of the essence when making the decision to elevate the threat level from possible to credible. The EPA has recommended a target time frame of two to eight hours to make the decision. This really depends on the situation and whether the potentially contaminated water will reach consumers or whether it can be isolated for further study. For example, potential contamination in a main trunk line may need a quicker response than two hours, while an isolated water storage tank that can be taken off-line while investigation is under way may make it acceptable to have a response time longer than eight hours.

All aspects of the possible threat should be considered, and the decision to raise the threat level from possible to credible is not one to be taken lightly. It is often at this point that actions to protect public health are implemented. These decisions may involve outside agencies, and at this point, the control of the incident may pass from the WUERM to an outside incident commander. Most possible threats will not make the transition to credible. Besides the information that was used to make the threat possible, the main categories of information to be considered, according to the EPA, are as follows:

(1) The results of site characterization, including observations from the site investigation as well as results from field safety screening and rapid field-testing should be considered. This information should include general information concerning the site as to its location and potential to cause harm. The general observations should also include a check for any physical evidence of tampering such as pumps or discarded containers. Also, the presence of outside indications of contamination such as dead animals in the area or unusual odors should be noted. The results of field analysis are crucial at this time. Due consideration as to the reliability or chances of false positives or negative for any of the field methods used should be duly noted. An inventory of the samples that were collected for further analysis and a log stating their disposition should be included with the field report. Also included should be the chain of custody for all samples as an actual incident will quickly transform into a criminal investigation. Samples should be sent to prearranged labs capable of performing the needed analysis.

(2) Summary information derived from an analysis of previous security incidents similar to the current threat warning. Information concerning all previous threats received by a utility should be well documented and catalogued for future reference. Comparing the current threat with past incidents of a similar nature can help in making the decision to terminate action or move it up the scale.

(3) Information from external sources that is relevant and available in a timely manner. There are a number of pertinent external sources of information and agencies that can offer help and useful information in determining the credibility of a threat. These sources include the state drinking water agency, the EPA, law enforcement agencies including the FBI, neighboring utilities, public health agencies, 911 call centers, homeland security warnings and alerts and, two specific tools that have been designed to help in these matters, the water ISAC and WCIT.[2]

> Water ISAC or the Water Information Sharing and Analysis Center is a Web-based resource that provides information on potential threats to the water industry. This is a secure site that is available only to those that qualify from the water industry. It is a national resource that provides information on a variety of water security-related subjects. It is available by subscription only and can be found at http://www.waterisac.org. Subscribers receive access to: timely email alerts about potential and actual physical or cyber attacks against drinking or wastewater systems; information on water security from the federal law enforcement, intelligence, public health and environment agencies; an extensive database of information on chemical, biological and radiological agents; notifications about cyber vulnerabilities and technical fixes; research, reports and other water security-related information; a highly secure means for quickly reporting incidents; vulnerability assessment tools and resources; guidance about emergency preparedness and response; the ability to participate in and review secure electronic forums on water security topics and helpful summaries of open-source security information.[6]
>
> WCIT or the Water Contamination Information Tool is a secure access restricted online database that provides information on contaminants of concern for water security. WCIT contains the most up-to-date information on water contaminants from peer reviewed sources and research. It includes data on potential contaminant names, availability, fate and transport, health effects and toxicity, medical information, drinking water treatment effectiveness, potential water quality and environmental indicators, sampling and analysis techniques, and helpful resource advise for utilities. All of this information can be extremely useful in helping first responders, including the utilities, make better informed decisions.[7]

Once the decision has been made that an incident represents a credible threat, public health response activities should be initiated. At this point, the objective of the public health response is to reduce or minimize the public's contact with the water supply, which has been deemed questionable in quality. The first and best option is always to contain a credible incident, preventing exposure to the public at large without interrupting service. If customers are located in and served by the containment area, further public health responses should be directed toward that populace; nevertheless, unless the containment is absolutely secure and isolation has been timely in its implementation, it is prudent to include the general populace in any warnings or alerts.

Unfortunately, containment is easier said than done. In cases where segregating and isolating the contaminated water is not an option, more drastic actions—such as an areawide alert to boil water, restriction of consumptive use, or an alert not to use the water for any purposes—may become necessary. It is important to consider the effects that any of these actions would have on consumers and the utility as a whole. Careful weighing of the level of credibility, the potential danger of the indicated attack, and the potential for a false alarm must be considered in making the decision. Note that consumers would bear some cost in lost revenue, health effects, and loss of confidence from any type of notification, regardless of whether the notification is the result of a false alarm or of an actual event.

What type of warning to issue, if any, will need to be decided on the basis of the information thus far obtained as to the likely threat. Does a warning to boil water adequately address the indicated contaminant, or is an order not to drink required? On detection of a credible threat, federal statute may mandate informing the public; under the Public Notification Rule,[8] in a "situation with significant potential to have serious adverse effects on human health as a result of short term exposure," notification of the public is required. If the situation also mandates that the public restrict its utilization of the water, it may become necessary to supply them with an alternative source of water for emergency purposes. Figure 8–12 illustrates this decision-making process.[2]

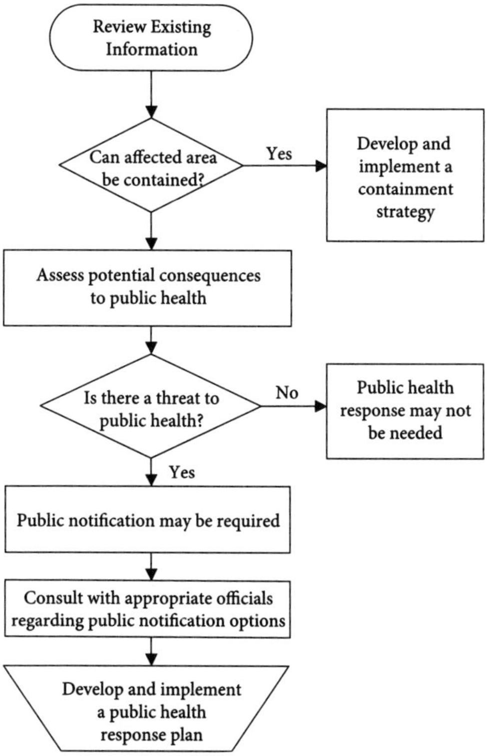

Fig. 8–12. Decision-making process for actions taken to protect public health in response to a credible contamination threat

Confirmed threats

The classification of an event as confirmed is based on sound evidence that the water supply has been contaminated. The problem is now no longer a threat but rather has become an incident. The most reliable source of information for this determination is confirmatory results obtained from an accredited laboratory. Further confirmation is possible through the Laboratory Response Network (LRN), which was established by the CDC to make these determinations. If laboratory confirmation is not possible, overwhelming evidence from such sources as the public health sector can be considered as confirmatory.

Unusual heath effects in the population or disease outbreaks as reported by local hospitals or 911 calls might indicate a potential contamination event. If there is good evidence to link these symptoms and occurrences to the water supply—especially when a potential water incident is under investigation—they may be considered confirmatory in nature.

At this stage, information about the specific contaminant being dealt with becomes particularly important. Proper identification of the contaminant is important in directing remediation efforts and advising healthcare workers. Time is of the essence in making these determinations; when samples are sent to the laboratory for confirmation after the credible determination, all pertinent data concerning tentative identification from online sensors and field studies should be included, so that analysis can be intelligently directed, speeding up the process. If the contaminant is unknown, WCIT may be able to narrow down the field.

Once an incident has reached the confirmed stage, it is doubtful that the utility will anymore be in charge of the response. State or federal incident control will most likely have taken command. An overview of the response is presented in figure 8–13.[2]

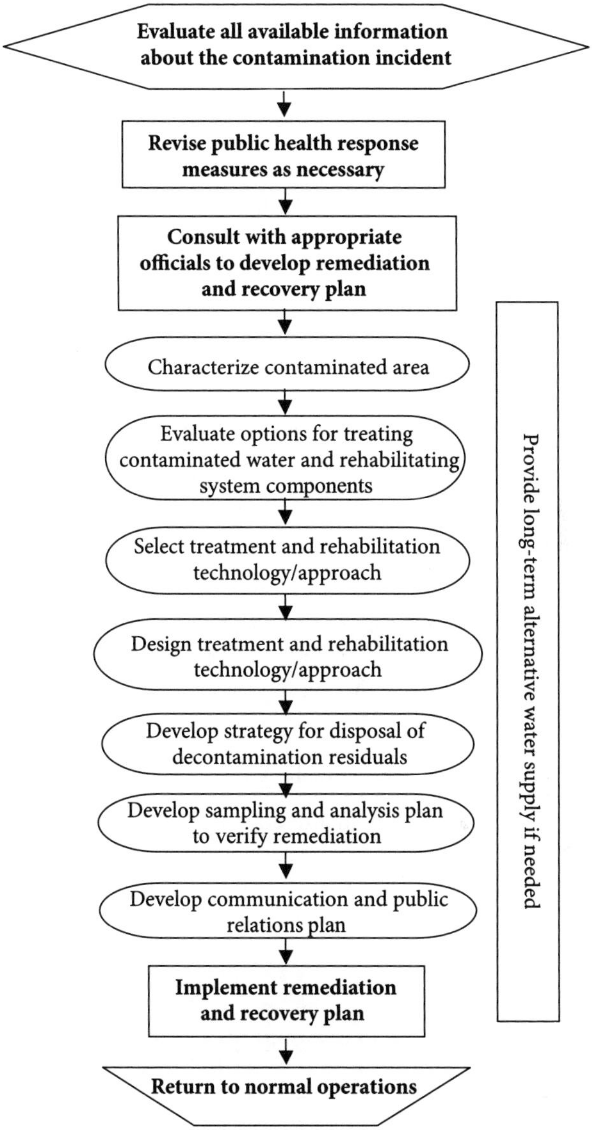

Fig. 8–13. Response to a confirmed contamination incident

Informing the Public

Even in combination with rapid confirmatory testing, the most accurate and sophisticated early-warning system becomes worthless without a quick and effective means for informing the public about a water contamination event. As mentioned earlier, in certain situations, the utility is required by statute to inform the public of questionable water quality. The rules require that the utility inform the public within 24 hours and make the utmost effort to see that all potential water users are informed. The warning must be designed to reach residential, transient, and nontransient users of the water system. The water systems are to use one or more of the following methods to warn the public:

- Broadcast media such as radio or television
- Posting of the warning in conspicuous locations throughout the area served by the system
- Hand delivery of the notice to persons served by the water system
- Another means approved in writing by the primary agency

Preapproval of alternative methods is prudent, because if an emergency does occur, the system will already be in place; valuable time will not be lost seeking approval. Alternative methods of informing the public include but are not limited to the following:

- Government-access TV channels
- Web sites (local government and other)
- Listserv e-mail
- Newspaper
- Telephone banks
- Broadcast faxes
- Mass distribution through community centers (churches, malls, restaurants)
- Door-to-door canvassing
- Town hall meetings
- Regular or special partner conference calls
- Reverse 911 messaging

In recent years, the popularity of reverse 911 systems has grown. These systems can be highly effective, but only if the phone numbers in the database are accurate and up to date. Hospitals and caregivers should always be informed of any warnings.[9]

In the wording and content of the warning, it is imperative to provide the public with as much information as is possible—beyond that a boil, do-not-drink, or do-not-use warning is in effect. The consumer should be provided with the reason for the warning, including information regarding what the suspected contaminant is, what the health effects and symptoms of exposure are, and steps they should take and why. A well-informed public is less likely to panic and fall prey to rumors that may be worse than the actual facts.

Cleaning up the mess

One of the final steps in addressing a water contamination emergency is dealing with cleanup. The difficulty of this endeavor depends entirely on which contaminant or mixture of contaminants was used. Contaminants such as the bacteria *E. coli* are easily removed through simple chlorination and flushing. Other compounds have proven to be extremely resistant to conventional remediation techniques.

As of this writing, a very incomplete understanding exists as to the fate and transport of many potential threat agents in the distribution system. Will the contaminant adhere to piping material? Is biofilm a factor? How are different types and ages of pipes affected by a given contaminant? Are the compounds degraded in water? Are the breakdown products toxic? Could some treatment regimes result in a more toxic state than before treatment?

In coordination with Hach HST, aggressive research programs sponsored by the American Water Works Research Foundation (AWWRF) and the U.S. Army Corps of Engineers are attempting to answer these questions. Preliminary research has indicated that some of the potential contaminants are indeed quite sticky, adhering to many types of piping. This research is important, in enhancing our understanding of what we will deal with in the event of an actual occurrence.

The cleanup cost of such an event could potentially be huge, if the agent is not readily removable from piping. Not only utility-owned piping but also interior plumbing may need to be replaced. If the agent is aerosolizable, whole areas—both inside and out—could need to be remediated. Lawn sprinklers could add to the problem by spreading toxic clouds of material over wide areas. Who would be financially responsible? These are all questions that will need to be addressed as the scientific knowledge is advanced.

One area of current research is the coupling of advanced algorithm-based monitoring techniques with an on-demand online treatment system. Online systems would be activated and controlled by the detection algorithms in the monitoring system. These devices have the potential, if properly designed, to protect a small area—such as a military base, a hospital, a school, or another icon facility. A number of different techniques to exploit this possibility are being explored. Even in the larger system as a whole, such a device could potentially lessen the damage done in an attack and reduce casualties and remediation costs.

Planning

The key to effectively responding to any incident is prior planning. Every step of the response should be carefully planned and coordinated. The Emergency Response Protocol Toolbox is a good start, offering helpful hints regarding planning. One point that warrants emphasis is it that all plans should have a level of redundancy and fallback as far as responsibilities are concerned. The type of emergency that we are dealing with has a high likelihood of resulting in casualties, and nothing precludes the possibility that key personnel orchestrating the response plan may be casualties themselves—hence the need for redundancy.

Furthermore, key assets outside the organization may not be available—either as a direct result of the attack or as the result of simultaneous attacks of another nature. This is why flexibility in planning is imperative. The ability of key personnel to adapt the basic plan to fit changing conditions or assets is crucial.

Drills and tabletop planning exercises can be employed to develop this ability. Exercises should be designed to rehearse more than the simple responses to various scenarios; they should also test responses to failures of various parts of the system to operate as conceived in the plan's inception. In the words of Publius Syrus, "It is a bad plan that admits of no modification."

Conclusion

Exactly how to respond in the case of a water contamination emergency presents a dilemma. The useful tools provided by the EPA and others are a staring point (and the majority of this chapter represents only a brief synopsis of these tools). No two emergency-response situations will play out in the same way. That is why we can expect nothing more than general guidance. However, this guidance, when combined with modern analytical tools currently available (and the even more advanced ones on the horizon), offers a powerful dynamic mechanism to direct our response.

The key to making the right choices has to do with knowledge and with the confidence we have in that knowledge. As the power of analytical techniques increases, the response scenario will change dramatically, and more of the response will become automated. If the online instruments say you have ricin in the system at 100 mg/L and we are 99.999% confident in the analytical results, then our response protocol becomes much more simple than if we were only 10% positive and still required extensive laboratory testing that could take hours or even days to reach the same confidence level. In the meantime, as the power of the analytical techniques slowly increases, we are forced to make the best decisions that we can with the information available. Let us hope that if an incident occurs, we will respond in an appropriate manner.

Notes

1. EPA. 2003–2004. Response protocol toolbox: Planning for and responding to drinking water contamination threats and incidents. http://cfpub.epa.gov/safewater/watersecurity/home.cfm?program_id=8#response_toolbox
2. Idem. Module 2: Contamination threat management guide.
3. Idem. Module 3: Site characterization and sampling guide.
4. ETV. http://www.epa.gov/etv/
5. SensIR. http://www.sensir.com?newsensir/Brochure/ExtractIR%20Product%20Note.pdf (accessed December 20, 2005).
6. Water ISAC. http://www.waterisac.org/services/default.asp (accessed January 3, 2006).
7. EPA. 2005. Water Contaminant Information Tool (WCIT) update. http://www.epa.gov/safewater/watersecurity/pubs/wcit_fs.pdf
8. Code of Federal Regulations 40, part 541, subpart Q.
9. EPA. 2003–2004. Response protocol toolbox: Planning for and responding to drinking water contamination threats and incidents. Module 5: Public health response guide. http://cfpub.epa.gov/safewater/watersecurity/home.cfm?program_id=8#response_toolbox

9

U.S. Water Utilities: Terrorism, Vulnerabilities, Legal Liabilities, and Protections Under the Safety Act[1]

Ours is an amazingly litigious society. We sue for everything and against everyone. We trip outside a store; we sue the store.

—Neil Cavuto

Introduction

Like it or not, we live in a litigious society. If we are unfortunate enough to suffer a terrorist attack on water supplies that results in casualties or property damage, you can guarantee it will wind up in the court system. Imagine a situation in which an attack takes place and the utility didn't do everything in its power to protect the public: lawsuit against the utility! Imagine if there was an attack and the utility had installed equipment to warn of and protect against an attack but it failed to function properly: lawsuit against the utility and the equipment manufacturer! Imagine if there was an attack and the utility had installed equipment to warn of and protect against an attack, and it functioned perfectly but the utility failed to respond appropriately: lawsuit against the utility, the city, the first responders, and the equipment manufacturer for failure to provide guidance! To help relieve this pressure of litigation and streamline the adoption of new antiterror technologies, the U.S. Congress drafted the Safety Act.

What Is the Safety Act?

The Safety Act, was enacted as part of the 2002 Homeland Security Act. The Safety Act was designed to promote the effective use of technologies that can be deployed by governments at the federal, state, and local levels and by private entities to guard against terror attacks on vulnerable infrastructure targets.[2] One such vulnerability that can be remediated by use of technology to make the nation safer is the nation's water distribution system.

By the close of 2004, however, the DHS had designated only four technologies for protection under the Safety Act, and a burdensome application process has discouraged applicants. Slow implementation of the Safety Act is diminishing its effectiveness and undermining its purpose. The Safety Act was intended to facilitate transactions by protecting not only manufacturers but also their customers, such as the water utilities that are increasingly concerned about how to best protect their assets from a terror attack.

Improved implementation by DHS would involve

- Faster technology deployment by water utilities to protect their distribution networks
- Legal protections to the water utilities for those actions
- Expedited action in securing associated legal protection in other areas where improved technology is needed, such as security for airlines and airports, seaports, nuclear facilities, and other infrastructure targets, such as water

Legal Liability Resulting from a Terror Attack

Lawsuits and compensation after 9/11

In the days after 9/11, Congress passed a law "to preserve the continued viability of the United States air transportation system" and to compensate the victims of the attacks.[3] Congress created an exclusive federal cause of action for damages related to the hijackings and crashes, and lawsuits for wrongful death, personal injury, property damage and business loss were consolidated in the Southern District of New York.[4] The law restricted the liability of the airlines to the limits of the liability insurance they maintained.[5]

The Safety Act adopts a similar federal cause of action and liability limit. The law of September 2001 also created the Victim Compensation Fund, an uncapped fund that was expected to provide more than one million dollars in awards to the families of each person killed.[6] Those who elected to participate in the Victim Compensation Fund were required to waive their right to bring lawsuits against the airlines and other defendants.[7]

Many lawsuits were filed nonetheless, pleading joint and several liability suits against many defendants. In the suits relating to the World Trade Center attacks, defendants include the manufacturer of the two hijacked aircrafts; the three airlines that carried the terrorists; and the many other airlines participating in the joint security system at the airports in Portland, Maine, and in Boston (Logan International Airport) through which the terrorists passed. Defendants also include the many security contractors at the airports and the City of Portland and the Massachusetts Port Authority, which operated the respective airports.[8]

Additional defendants were the Port Authority of New York and New Jersey, owner of the World Trade Center; the corporate lessees of the World Trade Center; and the corporations that designed the World Trade Center. At one point, more than a thousand suits by families were filed against the Port Authority, which itself lost employees as a result of the 9/11 attacks.[9,10] Among the allegations in these suits are claims of negligence, reckless conduct, conscious disregard for rights and safety, *res ipsa loquitur* (the common-law doctrine regarding injuries that do not occur in the absence of negligence), negligent infliction of emotional distress, punitive damages, and negligent selection of security contractors. Against the designers, the claims also include strict tort liability, negligent design, and breach of warranty. The defendants in these lawsuits filed motions to dismiss, and most of these motions were denied in September 2003, a decision that allowed the lawsuits to proceed.[11]

As of 2004, about a hundred families chose to pursue lawsuits instead of the Victim Compensation Fund. It was reported that "the flood of litigation is occurring despite the work of the . . . Fund, which Congress established, not only to help those directly harmed by the Sept. 11 plane crashes, but also to protect the airlines from lawsuits."[12] By the time the Victim Compensation Fund finished its work, about $7 billion in awards had been paid to families of those killed or injured.[13] Compensation from all sources to victims of the attacks—families of those killed, people who were injured, and businesses with property loss—has totaled $38.1 billion to date, of which $23.3 billion went to businesses. Insurance companies account for $19.6 billion of the total, government entities account for about $16 billion, and the remainder has gone to private charities.[14]

Municipal utility liability resulting from a backflow attack

In the legal aftermath of a backflow attack, as explained below, the entities to be sued would likely include the municipal or private water utility, or utilities providing services where the attack occurred, and the manufacturers and operators of any technology employed to protect the water system.[15] In addition to the hundreds of municipal water utilities around the country, there are many large private utilities, some of which are publicly traded, that serve the public and municipal customers, including 45 million people in the United States and Canada.

The majority of the U.S. population is served by community water systems that are publicly owned and operated. States generally benefit from sovereign immunity in state and federal courts, but the Supreme Court has noted that sovereign immunity "does not extend to suits prosecuted against a municipal corporation or other governmental entity which is not an arm of the State."[16] Most states have waived sovereign immunity in their own courts to varying degrees by Tort Claims Act statutes, as in New Jersey, Florida, Texas, California, Maryland, and Virginia, as well as the New York Court of Claims statute. These statutes are, in turn, subject to interpretation by the respective state common law, which often classifies state actions into categories of governmental and/or discretionary functions, which are immune, and proprietary[17] and/or nondiscretionary functions, which are not immune.[18] In general, municipal utilities benefit little from sovereign immunity.

Because many public water suppliers are quasi governmental, the defense against claims of liability has often been that, as public water suppliers, they are performing a governmental function and thus are insulated from liability. The courts have routinely rejected this view, holding that, with the exception of emergencies, public water suppliers are operating in a proprietary, rather than governmental, capacity.[19]

Municipal water utilities and operators would also have little protection in the event their water supplies are deliberately compromised. Commentators agree that municipal utilities could become liable following a terror attack. According to members of the American Bar Association Water Resources Committee, just as lawsuits followed the attacks on the World Trade Center in 1993 and 2001, similar lawsuits can be expected in the event that water supplies or infrastructure are sabotaged. For many water utilities, a large award could undermine their financial ability to continue providing needed services. Even a claim could affect a utility's bond rating.

Utilities could be sued under negligence theories; ironically, such legal actions might arise from attempts to make public water supplies more secure. For example, the EPA's recently issued guidelines detail the security measures water utilities are advised to implement immediately; if a particular utility fails to implement some or all of these measures or does so in a negligent manner, then the utility arguably should be liable for consequential damages. In the numerous jurisdictions to which comparative negligence applies, a utility theoretically could be held liable for some portion of the damages on evidence of minimal negligence.[20]

Furthermore, according to the Interim Voluntary Security Guidance for Water Utilities, released in December 2004, court rulings have found that a water utility must exercise reasonable care in operating and maintaining its system. The definition of reasonable care is key in determination of liability. As more water utilities implement security improvements, it could be argued that the definition of reasonable care is evolving to include installation of security systems that only a short time ago were rarely found in water systems.[21]

Liability of technology providers

Water utilities require equipment and technology provided by the private sector in order to improve security measures against a backflow or other terror attack. These companies are vulnerable to lawsuits even when their technology works as intended. Despite the massive resources devoted to the problem, no technology can prevent harm throughout every aspect of all the possible situations. In the event of an attack, lives can be saved with reliable technology deployed in an appropriate manner, but manufacturers expect lawsuits against them in the case of such an event, regardless of how their technology performs.

This litigious environment affects the choices companies make about whether it is worthwhile to develop antiterror technologies. One of the first companies to receive protection under the Safety Act has thus far declined to bring its product to market, because the insurance required is too expensive.

Most companies seeking protection have yet to receive it under the Safety Act. More than one year after the DHS issued an interim regulation on granting liability protections to antiterror technology manufacturers and six months since the first batch of companies to receive the protections were announced, only 4 out of more than 200 applications were approved. The law that was meant to encourage businesses to get into the homeland security market has apparently had little effect.[22]

Liability protection under the Safety Act

Should the hypothetical terror attacks described previously become a reality, a water utility could be sued for failure to take appropriate or recommended actions to protect its water distribution network. Even if it does take all feasible actions to prevent such an attack, the utility faces lawsuits over any technologies used to protect its network.

Consider the following scenario. In a major city, a contaminant-detection device alerts utility workers to a safety hazard involving a portion of the distribution network serving apartment buildings in the evening hours, after most of the city population has gone home.[23] The utility workers quickly take all necessary actions, which they have practiced, as required under government mandates for emergency preparedness. They alert appropriate utility personnel and government officials while investigating the hazard. They confirm the extent of the danger and relay the news to government officials, who initiate a civil emergency notification to the public over the broadcast networks. For the next several hours, newscasters urge the public not to touch their water at home, families and friends alert each other on cell phones, and utility officials take the steps necessary to protect the public.

Utility workers determine that the source of contamination is a deliberate attack, and the FBI and city police prepare for action. Urgent communications allow for precautions to be taken in other cities. The utility narrows down the possible locations of a backflow, and with the help of an alert public, the perpetrators are caught. Lives are saved, and by morning, it is announced to the public what steps need be taken to safely resume water use while further remediation is conducted. However, many injuries are reported, and some lives have been lost. Congress convenes and announces aid to the victims and their families.

If the technology employed by the utility is not officially designated under the Safety Act, as explained in the next section, the utility can be sued over its use and operation of the device, regardless of whether it fails or functions perfectly. If the technology employed is designated under the Safety Act, any suit relating to the performance of the device cannot be brought against the utility. Further, in the event that a suit relating to the device is allowed against the utility, the insurance maintained by the manufacturer is required to cover any liabilities of the utility. If the device functions as designed, the manufacturer also receives protection under the Safety Act. In the event that the device is defective and fails owing to negligence of a manufacturer, the courts would have the power to allow a remedy, which would be against the manufacturer alone and covered by its insurance.

Consider an alternative scenario where the detection device was never deployed. This could result in a hundredfold increase in the level of harm and injury.

Details of the Safety Act

The Safety Act statute

Congress enacted the Safety Act to promote the use of effective technology to combat terrorist threats and to offer manufacturers of such technology necessary protection from product-liability lawsuits[24,25]: "The Select Committee on Homeland Security believes that technological innovation is the Nation's front-line defense against the terrorist threat."[26] Part of the Homeland Security Act of 2002, the Safety Act is formally titled the Support Anti-Terrorism By Fostering Effective Technologies Act of 2002.[27] The Safety Act authorizes the Homeland Security Secretary to designate products and/or services as "qualified anti-terrorism technologies" (QATTs, or qualified ATTs) to receive certain protections.[28]

The DHS wrote regulations to implement the Safety Act, issuing a Proposed Rule on July 11, 2003, and an Interim Rule on October 16, 2003.[29] According to the Proposed Rule, the purpose of the Safety Act "is to ensure that the threat of liability does not deter potential manufacturers or sellers of anti-terrorism technologies from developing and commercializing technologies that could save lives."[30] There are two levels of protection for technologies under the Safety Act: designation as a qualified ATT; and at a higher level, certification as an approved product. Applications for both designation and certification are available on the DHS Safety Act Web site (www.safetyact.gov).

Benefits for designated antiterror technologies

Litigation management. The Safety Act establishes a litigation management framework and an exclusively federal cause of action for "claims arising out of . . . an act of terrorism when qualified anti-terrorism technologies have been deployed in defense against or response or recovery from such act and such claims result or may result in loss to the seller."[31] Among the protections for defendants under this framework, federal jurisdiction is exclusive, no punitive damages are allowed, noneconomic damages are limited (joint liability is prohibited and physical injury is required for such), and awards are reduced by collateral sources (e.g., insurance).[32] The federal cause of action applies to injuries related to technologies sold to the federal government and other customers (including state and local governments and commercial entities).[33] In other words, the Safety Act provides limited legal recourse to parties who believe they were injured by failure of designated technologies.

The appropriate federal district court is granted jurisdiction "over all actions for any claim for loss of property, personal injury, or death arising out of, relating to, or resulting from an act of terrorism when qualified ATTs have been deployed."[34]

Furthermore, "such Federal cause of action shall be brought only for claims for injuries that are proximately caused by sellers that provide QATT."[35] The DHS has set forth its view that under these provisions, "only one Federal cause of action exists for loss of property, personal injury, or death when a claim relates to performance or nonperformance of the seller's qualified and deployed anti-terrorism technology, and such cause of action may be brought only against the seller."[36] Thus, the Safety Act protects the seller's customers against the filing of lawsuits related to the performance of designated technologies.

Risk management. In addition, the Safety Act provides a risk-management framework, requiring sellers of qualified ATTs to obtain liability insurance against third-party claims in an amount, certified by the Homeland Security Secretary, that "will not unreasonably distort the sales price" of the technology.[37] Liability for all claims against a seller (including contribution and indemnity) "shall not be in an amount greater than the limits of liability insurance coverage required to be maintained by the seller under this section."[38] The DHS has discretion to determine the amount of liability insurance required to be maintained.[39]

These features of the Safety Act, creating a federal cause of action and establishing liability limits tied to insurance coverage, are analogous to provisions in the law that was enacted in September 2001 to protect the airline industry, following the disaster.[40] The Safety Act thus requires liability insurance and proportionate liability limits to protect the seller and provide recovery to plaintiffs.

The liability insurance required under this the risk-management framework will provide coverage against third-party claims arising from, relating to, or resulting from the sale or use of antiterror technologies.[41] The seller is required to enter into reciprocal waivers of claims with its contractors, subcontractors, suppliers, vendors, and customers, as well as with contractors and subcontractors of the customers.[42]

The Safety Act further protects the seller's customers in the event that any actions are allowed, related to the use and operation of technology, that are covered by the seller's insurance and liability limit.

Under the DHS interpretation, no lawsuits can be brought against customers for the performance or nonperformance of qualified ATTs. This interpretation should also prevent lawsuits related to the use or operation of qualified ATTs. However, should a lawsuit relating to the ATTs be allowed by a court, the customer is still protected by the seller's insurance and liability limits. Without Safety Act designation, customers could seek protection through their own insurance, but would not benefit from the liability limitation that exists under the Safety Act.

Government Contractor Defense for certified technologies

The litigation-management section of the Safety Act also provides for an additional benefit, the Government Contractor Defense (GCD), for technologies that are certified by DHS as approved products.[43] There "shall be a rebuttable presumption that the GCD applies" if a qualified ATT is subject to a product liability or other lawsuit; the presumption is overcome only in light of evidence that the seller acted fraudulently in submitting information to the Homeland Security Secretary. This defense applies to a sale of the product to the federal government or other customers (including state and local governments and commercial entities).

For certification as an approved product, additional steps by the applicant and the DHS are needed beyond designation as a qualified ATT: "Upon the seller's submission to the Secretary for approval of anti-terrorism technology, the Secretary will conduct a comprehensive review of the design of such technology and determine whether it will perform as intended, conforms to the seller's specifications, and is safe for use as intended."[44] Once the Secretary approves a technology, a certificate of conformance is issued, and the technology is placed on the Approved Product List for Homeland Security.[45]

The GCD is a common-law defense to tort liability established by the Supreme Court in a 1988 case, *Boyle v. United Technologies Corp.* In this ruling, the court held that a federal government contractor could not be held liable under state product-liability law for defects in military equipment when "(1) the United States approved reasonably precise specifications; (2) the equipment conformed to those specifications; and (3) the supplier warned the United States about the dangers in the use of the equipment that were known to the supplier but not to the United States." The Supreme Court concluded that "state law which holds Government contractors liable for design defects" presents a conflict with federal policy and "must be displaced."

In other words, the common law protects manufacturers from design-defect lawsuits as long as their designs have been approved by the federal government. The primary distinction codified by the Safety Act is that once a product is certified, the GCD applies regardless of whether the product is sold to the federal government, a state or local government, or even a private entity. This expands the scope of protection offered to manufacturers, to encourage the development of needed technologies and the sale of those technologies to the parties that need them most. In addition, sellers of technology need not design their technologies to government specifications in order for the GCD to apply to a certified technology under the Safety Act.[46] According to the DHS, "it is clear that any seller of an 'approved' technology cannot be held liable under the Act for design defects or failure to warn claims, unless the presumption of the defense is rebutted."[46]

The Safety Act leaves some questions about the statute's use of the common law unanswered.[47] For instance, in a situation in which common-law defenses would also be available to the contractor, would the statute's provision offer additional protection? Commentators have stated that the burden of proof under the statute should be more favorable to the manufacturer than under the common law, and the DHS has affirmed this view:

> The Department believes that Congress intended that, for purposes of applying the GCD, courts presume that all of the legal and factual requirements for establishment of the GCD by a government contractor are met by the existence of an applicable SAFETY Act Certification.[48]

The DHS has also stated its view that "Congress incorporated the Supreme Court's Boyle line of cases as it existed on the date of enactment of the Safety Act, rather than incorporating future developments...."[46] The common law has developed differences among the federal circuit court jurisdictions. For instance, many circuits have held that manufacturing defects are not covered by the GCD, but their rulings differ on this point. Were the common law of manufacturing defects to be interpreted under the Safety Act, courts would need to strike a balance between the purpose of the Safety Act, the intended protection of manufacturers, and the rights of injured victims in a case where a manufacturer was truly negligent. Thus, the Safety Act itself strikes an appropriate balance between these interests.

The Safety Act as the vehicle for government indemnification

On February 28, 2003, President Bush issued *Executive Order 13286*, granting responsibilities to the Homeland Security Secretary, amending previous executive orders, and limiting the previously existing ability of federal agencies to provide indemnification to government contractors, under *Public Law 85-804*, for "risks that the contract defines as unusually hazardous or nuclear in nature." The Defense Department is required to evaluate whether the Safety Act would be a more appropriate vehicle for indemnification—and determine that indemnification "is necessary for the timely and effective conduct of United States military or intelligence activities"—before granting indemnification through a contractual provision under *Public Law 85-804*.[49] Any other agency considering a contractor's request for indemnification for a product or service that could qualify under the Safety Act as a qualified ATT must first consult with the DHS and the Office of Management and Budget (OMB). The DHS must advise as to whether Safety Act protection would be more appropriate, and if the agency wishes to provide indemnification, it must receive approval from the OMB.

Executive Order 13286 affirmed the Safety Act as the proper and appropriate vehicle for the provision of liability protection to federal government contractors. Some major defense contractors have taken the position that they will not bid on certain contracts for antiterror services or products without Safety Act protections.

Implementation by the DHS

The Interim Rule delegates Safety Act responsibilities to the DHS undersecretary for science and technology.[50] By this rule, the DHS also created the application process for both designation and certification.[51] The certification process is largely the same as the designation process.[52]

Certification may be conditioned upon any specifications that the undersecretary finds appropriate.[53] Application kits are available on the DHS Web site. The DHS has also provided for pre-applications to be reviewed within 21 days, with feedback provided to assist businesses in how best to proceed. Within 90 days of receipt of a full application, the assistant secretary for plans, programs, and budget is to make a recommendation to the undersecretary that the application be approved or denied; or the assistant secretary may indicate that additional information is required and may extend this time period.[54] The undersecretary must likewise issue a decision, request additional information, or determine that additional time is necessary, within 30 days.[55]

In December 2004, the National Defense Industry Association (NDIA) held a two-day meeting to discuss implementation of the Safety Act. At that time, 153 pre-applications and 50 final applications had been submitted to the DHS; of the final applications, 2 had been denied, and only 4 had been approved. The technologies designated were Lockheed Martin's Risk Assessment Platform, which is a computer operating framework designed to conduct security risk assessment; Michael Stapleton Associates' SmartTech System, which is designed for screening of items for explosives and hazardous materials; Northrop Grumman's Biohazard Detection System, which analyzes mail for anthrax and other pathogens; and Teledyne Brown Engineering's Mobile Fluid Jet Access System, which is designed to cut through explosive devices with a high-powered water stream.[56]

The DHS had stated to Congress that it believes that, although the time needed to complete an application varies according to the size of the company involved, the average has been about 150 hours.[57] According to members of the NDIA, the time required in order to prepare a full application has been as much as 1,000 to 1,600 hours.[22] In December 2004, the DHS announced that it had prepared a draft revision of the application kit and sought additional comments on the application process.[58]

The chairs of the House Judiciary Committee, the Committee on Government Reform, and the Select Committee on Homeland Security wrote to Homeland Security Secretary Tom Ridge in May 2004, to relay their concerns about the implementation of the Safety Act.[59] The chairs wrote,

> As the statute itself suggests, the analysis to be undertaken by the Department for the designation and certification of a given technology was intended to be simple and straightforward—a means of *facilitating* transactions, *not* erecting additional barriers to deployment. . . . It is absolutely essential that the Department initiate a process to prioritize applications for Safety Act designation and certification, and ensure that critical technologies receive expedited treatment.

Further, the chairs did not view this process as "requiring the Department to insert itself in a pending transaction for the purpose of establishing performance standards for a given technology. . . ." The letter continued,

> If, for example, a city government and an anti-terrorism technology manufacturer have negotiated a contract to purchase biohazard detectors, and have made consummation of the deal contingent upon Safety Act designation and certification of the biohazard detectors, the Department's review should not involve a de novo determination of whether the detectors meet a particular performance standard. . . . Unless the design or operation of the product itself poses inherent risks to the public, the technology should be promptly designated or certified. . . . Where pending procurements are involved, the Department should defer to the judgment of the buyer and utilize information already provided in connection with the procurement, rather than reconstruct a process the parties already have diligently undertaken.

The DHS replied in June 2004 in a letter from the undersecretary for science and technology, Charles E. McQueen.[57] Mr. McQueen wrote, "While I understand your concerns and your desire for a more streamlined process, I believe the process we have implemented is consistent with the minimum requirements of the Act." Mr. McQueen agreed with some specific points raised by the congressional committee chairs and disagreed with others, such as the role of DHS when a transaction is pending between a municipal government and technology manufacturer. This issue has come to the forefront, because today municipalities are responding to terror threats with the help of private technology partners. These parties do not yet benefit from protections under the Safety Act, as Congress intended.

In addition, procurements are underway for bids to be contingent on protection under the Safety Act. Uncertainty over the status of an antiterror technology consequently affects the willingness of parties to enter into transactions related to those technologies. Future DHS policies may address these procurement issues.

Conclusion

The future of the Safety Act, like the nature of terror threats to come, is uncertain, but the risks of terror threats remain. The protection of water distribution systems is just one area where needed improvements cannot occur without the deployment of new technology. Municipal entities and their private partners have a duty to improve security and face liability if they fail to take appropriate measures recommended by the industry. However, providers of technology face liability simply by offering the needed protection, and many will not provide it without the application of the Safety Act.

If the Safety Act functions as Congress intended, it will encourage the deployment of necessary technologies designed to improve national security. In order for this to occur, the DHS must do more to facilitate the assessment of applications for designation and certification under the Safety Act. The provisions afforded by the Safety Act protect not only manufacturers but also the very entities responsible for infrastructure security: federal agencies, states, municipalities, and their private partners. With maximum implementation by the DHS, the Safety Act will facilitate their efforts to protect the public with the most robust security measures that technology can provide.[60]

Notes

1. This chapter was prepared with the help of Richard J. Conway and Joseph R. Berger of Dickstein, Shapiro, Morin & Oshinsky, LLP, Washington, DC, in January 2005. It is not meant to be taken as definitive legal advice but as a general guideline in assessing potential liabilities and understanding the role that the Safety Act plays in these considerations.

2. The Safety Act is codified at U.S. Code 6, sections 441–444, with further implementation at Code of Federal Regulations 6, sections 25.1–25.9, and is administered by the DHS (for more information, see http://www.safetyact.gov/).

3. The Air Transportation Safety and System Stabilization Act (ATSSSA), 2001. Public Law 107-42, 115 Statute 230 (September 22).

4. ATSSSA, section 408(b)(1). See http://nysd.uscourts.gov/Sept11Litigation.htm. In addition to the victims' families, the owners and insurers of property in the World Trade Center and surrounding area filed a separate complaint (WTC property damage plaintiffs file master complaint. 2003. *Andrews Aviation Litigation Reporter.* January 21.) A suit seeking $5 billion was also filed against both manufacturer Motorola and the City of New York, over handheld radios used by firefighters, alleging product design defect, failure to warn, and fraudulent misrepresentation; more than a year later, this suit was dismissed because the plaintiffs also participated in the Victim Compensation Fund (9/11 Firefighters' families sue Motorola, but court finds they waived right to sue. 2004. *Product Safety & Liability Reporter.* Vol. 32, issue 14. April 5). In another consolidated suit, more than a thousand workers involved in the rescue and

cleanup efforts following 9/11 sued the City of New York over the adequacy of safety equipment (Julius A. Rousseau III et al. 2004. Scope of Southern District of New York's jurisdiction for claims arising from September 11, 2001. *Mealey's Litigation Report: Insurance.* April 13). A similar lawsuit by recovery workers over safety equipment was later filed against the leaseholder of the World Trade Center and the companies that supervised the cleanup. (Ground Zero workers file billion-dollar suit against WTC owners. 2004. *Andrews Toxic Torts Litigation Reporter.* November 23).

5. ATSSSA, section 408(a)(1). For a review by the plaintiffs' lawyers in these 9/11 suits, see Kreindler, James P., and Brian J. Alexander. 2004. September 11 aftermath: A perspective on the VCF and litigation. *Air & Space Lawyer.* 18 (1): 17–21.

6. Henriques, Diana B., and David Barstow. 2001. A nation challenged—victims' compensation: Fund for victims' families already proves sore point. *New York Times.* October 1.

7. Two years later, the average award for loss of life was $1.6 million, and less than half of victim's families had applied to the Victim Compensation Fund (Henriques, Diana B. 2003. Concern growing as families bypass 9/11 victims' fund. *New York Times.* August 31). Meanwhile, many lawsuits against the airlines were filed as statute of limitation deadlines approached (Weiser, Benjamin. 2003. Two years later—lawsuits: Families of victims file to meet a legal deadline. *New York Times.* September 11). Many plaintiffs ultimately abandoned their lawsuits as the deadline (December 22, 2003) for participation in the Victim Compensation Fund approached. Chen, David W. 2003. Applicants rush to meet deadline for Sept. 11 fund. *NewYork Times.* December 23.

8. Plaintiffs' Amended Flight 11 Master Liability Complaint; Plaintiffs' Amended Flight 175 Master Liability Complaint, *In re September 11, 2001 Litigation,* No. 21 MC 97 (AKH) (State District of New York).

9. Sept. 11 anniversary marks lawsuit deadline for Port Authority. 2002. *Best Wire.* September 12.

10. Chen, David W. 2002. Suits by 950 families allege safety lapses at the Towers. *New York Times.* September 14.

11. *In re September 11, 2001 Litigation.* 2003. 280 Federal Superior Court. 2d 279 (State District of New York). See Henreiques, Diana B., and Susan Saulny. 2003. Two years later—lawsuits: Judge's ruling opens door for more families to sue airlines and Port Authority. *New York Times.* September 10.

12. Eaton, Leslie. 2004. Lingering 9/11 anger finds its outlet in courts. *International Herald Tribune.* September 10.

13. Chen, David W. 2004. Striking details in final report on 9/11 fund. *New York Times.* November 18.

14. Chen, David W. 2004. New study puts Sept. 11 payout at $38 billion. *New York Times.* November 9.

15. Dinh, Mary, and Seth Drewry. 2004. The SAFETY Act's impact on US homeland security. Janes.com. July 20. "For example, a manufacturer of a chemical or biological agent detection device could face one or several class action suits worth billions of dollars if the equipment they produced failed to prevent an attack using one of these agents. This prospect has made government contractors . . . cautious about the programmes in which they choose to participate."

16. *Alden v. Maine*. 1999. 527 U.S. 706, 756.

17. Meaning an activity conducted for the benefit of the municipality as a corporate entity, rather than for the benefit of the public. "When a municipality is in the business of selling water to customers for profit or revenue, it is engaged in a proprietary function" (*Junior College District of St. Louis v. City of St. Louis*. 2004. No. SC 85583, 2004 WL 2663621 (Mo.) (en banc). Even a state can sue a municipality over negligence in the maintenance of municipal water lines; e.g., see *State of Texas v. City of Galveston*. 2004. No. 01-03-00557-CV, 2004 WL 2066448 (Tex. App.).

18. New York maintains the "general rule that a municipality may be liable if its agents are acting in a proprietary capacity, but not, absent some special undertaking, in a governmental capacity" (Hancock, Stewart F., Jr. 1993. Municipal liability through a judge's eyes. 44 Syracuse L. Rev. 925, 936). In Florida, "several decades of Florida Supreme Court decisions construing Florida's waiver statute have generated a body of case law [such that] there are no defined legal boundaries of governmental tort liability and there is no clear framework with which to analyze immunity" (Bustin, Thomas A., and William N. Drake, Jr. 2003. Judicial tort reform: Transforming Florida's waiver of sovereign immunity statute. 32 Stetson L. Rev. 469, 469-470). In Texas, a municipality is by statute liable "for damages arising from its governmental functions . . . including . . . health and sanitation services" (Shaw, Tara L. 2002. Is Texas waiving good-bye to sovereign immunity? 3 Tex. Tech. J. Tex. Admin. L. 225, 232.

19. Meyland, Sarah J. 1993. Land use & the protection of drinking water supplies. In *Pace Environmental Law Revue*. Vol. 10: 563, 587. In the context of ordinary utility activities, "there is agreement . . . that the doctrine of sovereign immunity affords no protection to a municipal water distributor guilty of negligence in regard to the escape of water from its service pipes," and "it has been held that a water distributor may incur liability for negligently maintaining service pipes in such a condition that impure water is furnished to the consumer" (Water distributor's liability for injury due to condition of service lines, meters, and the like, which service individual consumer. 2004. 20 A.L.R.3d 1363, section 2).

20. De Young, Tim, and Adam Gravley. 2002. Coordinating efforts to secure American public water supplies. 16-WTR Nat. *Resources & Environment* 146, 152. "Some water utility officials believe that the leading threat to the nation's water supply may be the use of backflow pressure to introduce poisons into local water distribution systems."

21. American Water Works Association, American Society of Civil Engineers, and Water Environment Foundation. 2004. Water infrastructure security enhancements: Interim voluntary security guidance for water utilities. http://www.awwa.org/science/wise/#P7_623. "Once a vulnerability assessment is complete, the resulting recommendations . . . could be considered as notice of a dangerous condition. This notice could potentially result in liability if the recommendations are not addressed."

22. Starks, Tim. 2005. Best laid plans: Effort to lure homeland business with liability protection falls far short of goals. *Congressional Quarterly Homeland Security*. January 7. "And for companies that have not won the protections of the Safety Act yet, industry officials said, the best work to describe their view of the law so far is frustrated. Many that have applied have found the process maddening, and some companies have not even bothered to apply because of the expectation that it is not worth the trouble." Insurance costs for another company to win Safety Act designation, a small security firm, have ballooned from $5,000 per year prior to September 11, 2001 to more than $500,000 (Block, Robert, and J. Lynn Lunsford. 2004. U.S. gives liability protection to anti-terror firms. *The Wall Street Journal*. June 18).

23. EPA has assisted utilities in preparation of emergency response plans for similar scenarios by issuing the Response protocol toolbox: Planning for and responding to drinking water contamination threats and incidents. "Utility staff possess an extensive knowledge about the physical configuration, operation, and water quality of their system. This knowledge should be utilized throughout the entire threat evaluation process. . . . Furthermore, during advanced stages of an incident, the understanding of distribution system hydraulics by operations staff and engineers will be critical to the rapid assessment of the propagation of a suspected contaminant through a system" (Module 2: Contamination threat management guide).

24. Homeland Security Act of 2002, Public Law 107-296, 16 Statute 2135.

25. House of Representatives. 2002. Report no. 107-609(I), reprinted in *U.S. Code Congressional and Administrative News.* 1352, 1399.

26. Ibid. "Unfortunately, the Nation's products liability system threatens to keep important new technologies from the market where they could protect our citizens. In order to ensure that these important technologies are available, the Select Committee believes that it is important to adopt a narrow set of liability protections for manufacturers of these important technologies."

27. U.S. Code 6, section 441–444. In particular, see part G of title 6, which is the Domestic Security Title of the U.S. Code.

28. U.S. Code 6, section 441(b): QATT status is designated according to listed criteria including magnitude of risk exposure to the public if technology is not deployed and likelihood that the technology would not be deployed without the protections of the Safety Act. Also, referring to U.S. Code 6, section 441(l): QATT is defined to include "any product, equipment, service . . . device, or technology . . . designed, developed, modified, or procured for the specific purpose of preventing, detecting, identifying, or deterring acts of terrorism or limiting the harm such acts might otherwise cause, that is designated as such by the Secretary."

29. Regulations implementing the Safety Act: Proposed Rule, 68 Federal Register, at 41,420 (July 11, 2003); Interim Rule, 68 Federal Register, at 59,648 (October 16, 2003) (codified at Code of Federal Regulations 6, part 25).

30. Proposed Rule, 68 Federal Register, at 41,420.

31. U.S. Code, section 442(a)(1).

32. U.S. Code 6, section 442(b).

33. The substantive law is that of the state where the act of terrorism occurs, except where preempted by federal law; see section 442(a)(1). The term "non-federal government customers" is defined in section 444(6).

34. U.S. Code 6, section 442(a)(2).

35. U.S. Code 6, section 442(a)(1).

36. Preamble to Proposed Rule, 68 Federal Register, at 41,420 and 41,423. This is "the best reading" of these provisions and "the reading the Department is inclined to adopt." Under Supreme Court case law, federal courts give deference to an agency's reasonable interpretation of a statute or its own regulations.

37. U.S. Code 6, section 443(a)(1),(2).
38. U.S. Code 6, section 443(c).
39. Code of Federal Regulations 6, section 25.4(h).
40. With the Safety Act, "Congress balanced the need to provide recovery to plaintiffs against the need to ensure adequate deployment of anti-terrorism technologies by creating a cause of action that provides a certain level of recovery against sellers, while at the same time protecting others in the supply chain" (Interim Rule, 68 Federal Register, at 59,693).
41. U.S. Code 6, section 443(a)(3),(4).
42. U.S. Code 6, section 443(b).
43. U.S. Code 6, section 442(d)(1).
44. U.S. Code 6, section 442(d)(2).
45. U.S. Code 6, section 442(d)(3).
46. Proposed Rule, 68 Federal Register, at 41,422.
47. Levin, Alison M. 2004. The Safety Act of 2003: Implications for the Government Contractor Defense. 34 Pub. Cont. L.J. 175.
48. Interim Rule, 68 Federal Register, at 59,687.
49. Executive Order no. 13,286. 2003. Section 73. February 28.
50. Code of Federal Regulations 6, section 25.2.
51. Code of Federal Regulations 6, sections 25.3 and 25.5 (designation) and section 25.7 (certification).
52. Code of Federal Regulations 6, section 25.7(a). A certification application cannot be filed without a designation application, and DHS may not issue a certification without a designation.
53. Code of Federal Regulations 6, section 25.7(g).
54. Code of Federal Regulations 6, sections 25.5(d) and 27.5(d).
55. Code of Federal Regulations 6, sections 25.5(e) and 27.5(e).
56. Department of Homeland Security announces first designations and certifications under the Safety Act. 2004. Office of the Press Secretary, Department of Homeland Security. June 18.
57. Charles E. McQueen. 2004. Letter to Christopher Cox, June 14.
58. Revision of currently approved information collection requests for SAFETY Act of 2002. 2004. 69 Federal Register. 72207. December 13.
59. Sensenbrenner, James, Jr., Christopher Cox, and Tom Davis. 2004. Letter to Tom Ridge. May 11.
60. Since the preparation of this manuscript, DHS has streamlined the review process in an attempt to facilitate the process for new technologies.

10

Challenges Ahead

There is a Chinese curse, which says, "May he live in interesting times." Like it or not, we live in interesting times.

—Robert F. Kennedy

These are indeed interesting times for the water industry. The many challenges of aging infrastructure, increasing water demand, limited water supplies, reduced funding, accidental contamination, and pollution have been exacerbated by the new threat of intentional attack. Beset on all sides, the water industry has experienced increasing difficulty in focusing energy and funding on all the areas that require or deserve attention. It becomes easy to ignore or downplay challenges that are not immediate in nature. All of our energy goes to putting out fires, rather than long-term planning.

It has been over four and a half years since the last deadly attack occurred on American soil. While we will never forget the attacks of 9/11, the awfulness of that day has begun to lose some of its sharpness as time flows inexorably onward. This has led to a sense of complacency about security in general—and about the security of our water systems in particular. A common opinion is that if the danger of such attacks has not passed, such attacks are at least not likely to occur anytime soon.

The reasoning behind this assumption of safety in the short term is that if al Qaeda or other groups had plans to attack our water supplies, then they would have done so by now. However, because the long operational view of terrorist groups is not compatible with a Western mind-set or time frame, it is not always acknowledged by our culture. Planning for large-scale events can take years. For example, al Qaeda spent years planning the 9/11 attacks.

Even longer-term strategic planning that is generational in nature is common. Sleeper cells can be inactive for years or even decades before being called into action. A terrorist manual entitled *Encyclopaedia of the Afghani Jihad,* multiple copies of which were discovered in a raid on accused terrorist Abu Hamza's house in Britain, recommends that agents should be sent to any country intended as a target at least 10 years before jihad begins.[1] Thus, a goal is set even though those doing the planning know that they will not be around to see their plans executed. It is imperative that we recognize this concept of long time frames when formulating our own plans for defensive action.

From a defensive standpoint, the long planning schedules utilized by our potential enemies can be either a blessing or a curse. On the one hand, the long lag between attacks can result in a lowering of the guard and, hence, an increased vulnerability. On the other hand, when a strategic objective has not been adequately protected in the past, such as water, the delay between events offers the possibility of hardening the target so as to discourage any potential threats.

Terrorists have had a long and documented interest in attacking water supplies. There is little doubt that our water supplies are vulnerable to either deliberate or accidental disruption. Of the myriad possible ways that the water system could be attacked, the most frightening attack mode, with the greatest potential to cause mass casualties, would be an intentional contamination event. Of all the areas where contamination could occur, the distribution system is most at risk. The potential damage that such an attack could cause puts it in the same class as a WMD. The ease of mounting such an attack should indeed frighten us.

Whether it comes from international terrorists, disgruntled employees, or some other radical group, it is inevitable that an attack will occur in the future. This book has outlined the steps that we can take to thwart such an attack or at least reduce the severity of its impact. Infrastructure hardening through physical security, increased emphasis on cybersecurity, monitoring for water contamination events, and detailed planning of responses to any such attack all decrease our vulnerability. The secret is not to become complacent just because such an attack has not as yet been successfully orchestrated. It is imperative that we make use of the lull that has been afforded us to aggressively address the current vulnerability of our water systems.

If this is done in an intelligent manner, there is no need for the burden placed on our finances and time to be unbearable. Intelligent design of everything from basic infrastructure to cybersecurity and monitoring platforms can streamline existing processes and improve the quality of the water reaching end users. Hopefully, this book not only has been effective in pointing out the vulnerabilities that plague our systems but has offered potential solutions to these problems that will result in a safer water supply for everyone.

Advances in technology will present new opportunities to make our water increasingly safe and secure. However, the state of the art today offers significant possibilities not only for increased security but also for streamlining of efficiencies in operations and delivery of high-quality safe water to everyone. It is up to us whether to take advantage of the attentiveness currently focused on our water supplies in light of the threat of terrorism; the alternative is to ignore the threat and fail to implement the improvements possible using currently available technologies. Let us hope that we make the wise choice, improving our current system such as to lessen vulnerability—while improving water quality and safety in general—before another disaster strikes; otherwise, we will be left to wonder why we didn't use our common sense and implement these safeguards in time.

Notes

1. Woolcock, Nicola. 2006. Blueprint for terrorism found in house." *The Sunday Times.* January 12. http://www.timesonline.co.uk/article/0,,2-1981588,00.html

Appendix A: Chemical and Biological Agents of Concern

Chemical Agents on the CDC List of Concern

Details on each of the chemical agents listed below can be found online (at http://www.bt.cdc.gov/agent/agentlistchem.asp).

Abrin
Acids (caustics)
Adamsite (DM)
Ammonia
Arsenic
Arsine (SA)
Barium
Benzene
Biotoxins
Blistering agents (vesicants)
Blood agents
Brevetoxin
Bromine
Bromobenzyl cyanide (CA)
BZ
Carbon monoxide
Caustics (acids)
Chlorine (Cl)
Chloroacetophenone (CN)
Chlorobenzylidenemalonontrile (CS),
Chloropicrin (PS),
Choking/lung/pulmonary agents
Colchicine
Cyanide
Cyanogen chloride (CK)
Dibenzoxazepine (CR)
Digitalis
Diphosgene (DP)
Distilled mustard (HD)
Ethylene glycol
Fentanyls and other opioids
Hydrochloric acid (hydrogen chloride)
Hydrofluoric acid (hydrogen fluoride)
Hydrogen chloride (hydrochloric acid)
Hydrogen cyanide (AC)
Hydrogen fluoride (hydrofluoric acid)
Incapacitating agents
Lewisite (L, L-1, L-2, L-3),
Long-acting anticoagulant (super warfarin)
Lung/choking/pulmonary agents
Mercury
Metals
Methyl bromide
Methyl isocyanate
Mustard gas (H) (sulfur mustard)
Mustard/lewisite (HL)
Mustard/T
Nerve agents
Nicotine
Nitrogen mustard (HN-1, HN-2, HN-3)
Opioids
Organic solvents
Osmium tetroxide
Paraquat
Phosgene (CG)
Phosgene oxime (CX)
Phosphine
Phosphorus, elemental, white or yellow
Potassium cyanide (KCN)
Pulmonary/choking/lung agents
Ricin
Riot-control agents/tear gas
Sarin (GB)
Saxitoxin
Sesquimustard
Sodium azide
Sodium cyanide (NaCN)
Sodium monofluoroacetate (compound 1080)
Soman (GD)
Stibine
Strychnine
Sulfur mustard (H) (mustard gas)
Sulfuryl fluoride
Super warfarin (long-acting anticoagulant)
Tabun (GA)
Tear gas/riot-control agents
Tetrodotoxin
Thallium
Toxic alcohols
Trichothecene
Vesicants (blistering agents)
Vomiting agents
VX
White phosphorus

Biological Agents on the CDC List of Concern

Details on each of the biological agents listed below can be found online (at http://www.bt.cdc.gov/agent/agentlist.asp).

Anthrax (*Bacillus anthracis*)
Arenaviruses

Bacillus anthracis (anthrax)
Botulism (*Clostridium botulinum* toxin)
Brucella species (brucellosis)
Brucellosis (*Brucella* species)
Burkholderia mallei (glanders)
Burkholderia pseudomallei (melioidosis)

Chlamydia psittaci (psittacosis)
Cholera (*Vibrio cholerae*)
Clostridium botulinum toxin (botulism)
Clostridium perfringens (Epsilon toxin)
Coxiella burnetii (Q fever)

Ebola virus hemorrhagic fever
E. coli O157:H7 (*Escherichia coli*)
Emerging infectious diseases (e.g., Nipah virus and hantavirus, Epsilon toxin of *Clostridium perfringens*)
Escherichia coli O157:H7 (*E. coli*)

Food safety threats (e.g., *Salmonella* species, *Escherichia coli* O157:H7, *Shigella*)
Francisella tularensis (tularemia)

Glanders (*Burkholderia mallei*)

Lassa fever

Marburg virus hemorrhagic fever
Melioidosis (*Burkholderia pseudomallei*)

Plague (*Yersinia pestis*)
Psittacosis (*Chlamydia psittaci*)

Q fever (*Coxiella burnetii*)
Ricin toxin (from *Ricinus communis* [castor beans])
Rickettsia prowazekii (typhus fever)

Salmonella species (salmonellosis)
Salmonella typhi (typhoid fever)
Salmonellosis (*Salmonella* species)
Shigella (shigellosis)
Shigellosis (*Shigella*)
Smallpox (variola major)
Staphylococcal enterotoxin B

Tularemia (*Francisella tularensis*)
Typhoid fever (*Salmonella typhi*)
Typhus (*Rickettsia prowazekii*)

Variola major (smallpox)
Vibrio cholerae (cholera)
Viral encephalitis (alphaviruses [e.g., Venezuelan equine encephalitis, eastern equine encephalitis, western equine encephalitis])
Viral hemorrhagic fevers (filoviruses [e.g., Ebola, Marburg] and arenaviruses [e.g., Lassa, Machupo])

Water-safety threats (e.g., *Vibrio cholerae, Cryptosporidium parvum*)

Yersinia pestis (plague)

Military List of Chemicals of Concern in Water

The compounds in this list are from the May 1999 guidelines on short-term chemical exposure for deployed military personnel (U.S. Army Center for Health Promotion and Preventive Medicine).

Acenaphthene
Acenaphthylene
Acetone
Acifluorfen
Acrylamide
Acrylonitrile
Adipate (diethylhexyl)
Alachlor
Aldrin
Ametryn
Ammonia
Ammonium sulfamate
Anthracene
Antimony
Aroclor-1016
Aroclor-1254
Arsenic
Atrazine

Baygon
Bentazon
Benzene
Benzo[*a*]anthracene
Benzo[*b*]fluoranthene
Benzo[*k*]fluoranthene
Benzo[*a*]pyrene
Beryllium
Bis(2-ethylhexyl)phthalate
[Di-(2-ethylhexyl)phthalate]
Boron
Bromacil
Bromochloromethane
Bromodichloromethane
Bromoform
Bromomethane
Butylate
sec-Butylbenzene
BZ

Cadmium
Carbaryl
Carbofuran
Carbon disulfide
Carbon tetrachloride
Chloral hydrate
Chloramben
Chlordane
Chloride
Chlorobenzene
Chlorodibromomethane
Chloroform
Chloroisopropyl ether
Chloromethane
Chlorophenol
Chlorothalonil
o-Chlorotoluene
p-Chlorotoluene
Chlorpyrifos
Chromium (total)
Chromium(III)
Chromium(VI)
Chrysene
Copper
Cumene
Cyanazine
Cyanide

2,4-D [Dichlorophenoxyacetic acid]
Dalapon
DCPA [Dacthal]
DDT
Diazinon
Dibromoacetonitrile
Dibromochloromethane
Dibromochloropropane
Dicamba
Dichloroacetic acid,
Dichloroacetonitrile
m-Dichlorobenzene
o-Dichlorobenzene
p-Dichlorobenzene

Dichlorodifluoromethane
1,2-Dichloroethane
1,1-Dichloroethylene
cis-1,2-Dichloroethylene
trans-1,2-Dichloroethylene
Dichloromethane [Methylene chloride]
2,4-Dichlorophenol
1,2-Dichloropropane
1,3-Dichloropropene
Dieldrin
Di(2-ethylhexyl)phthalate
  [Bis(2-ethylhexyl)phthalate]
Diisopropyl methyl phosphonate
Dimethrin
Dimethyl methyl phosphonate
1,3-Dinitrobenzene
2,4-Dinitrotoluene
2,6-Dinitrotoluene
Dinoseb
1,4-Dioxane
Diphenylamine
Disulfoton
Dithiane
Diuron

EA 2192
Endothall
Endrin
Epichlorohydrin
Ethylbenzene
Ethylene dibromide
Ethylene glycol
ETU [Ethylene thiourea]

Fenamiphos
Fluometron
Fluoranthene
Fluorene
Fluorotrichloromethane
Fonofos
Formaldehyde

GA [Tabun]
GB [Sarin]
GD [Soman]
Glyphosate

Heptachlor
Heptachlor epoxide
Hexachlorobenzene
Hexachlorobutadiene
Hexachloroethane
n-Hexane
Hexazinone
HMX

Isophorone
Isopropyl methyl phosphonate

Lead compounds
Lewisite
Lindane

Magnesium
Malathion
Maleic hydrazide
MCPA
Mercury (inorganic)
Mercury (methyl)
Methomyl
Methoxychlor
Methyl ethyl ketone
Methyl *tert*-butyl ether [MTBE]
Methyl parathion
Metolachlor
Metribuzin
Molybdenum
Molybdenum trioxide

Naphthalene
Nickel
Nitroguanidine
p-Nitrophenol

Oxamyl [Vydate]

Paraquat
Phenanthrene
Pentachlorophenol
Phenol
Picloram
Prometon
Pronamide
Propachlor
Propazine
Propham
n-Propylbenzene
Pyrene

RDX

Silver
Simazine
Strontium
Styrene
Sulfate
Sulfur mustard [HD]

T-2 toxin
TCDD
Tebuthiuron
Terbacil
Terbufos
Tetrachlorodibenzodioxin
1,1,1,2-Tetrachloroethane
Tetrachloroethylene
Thallium
Toluene
Toxaphene
2,4,5-TP
Trichloroacetic acid
Trichloroacetonitrile
Trichlorobenzene
1,1,1-Trichloroethane
1,1,2-Trichloroethane
Trichloroethylene
Trichlorophenoxyacetic acid
Trichloropropane
Trifluralin
1,2,4-Trimethylbenzene
1,3,5-Trimethylbenzene
Trinitroglycerol
2,4,6-Trinitrotoluene

Vanadium
Vinyl chloride
VX

Xylenes

Zinc
Zinc chloride

Appendix B: Suspicious Incident Information Reporting Form

Suspicious Incident Information Reporting Form

General Information	
YOUR NAME/CONTACT #	
Date and Time of Incident	
Location (Address or Street)	
Weather Conditions	

Details of Incident

Suspicious Person #1 Physical Description

Gender M / F Ethnicity ____________ Height______ Weight______

Unique Physical Characteristics____________

Clothing Type/Style/Color____________

Vehicle, if any Year/Make/Model____________ Color____________

License Plate Information: State____________ Plate #____________

Suspicious Person #2 Physical Description

Gender M / F Ethnicity ____________ Height______ Weight______

Unique Physical Characteristics____________

Clothing Type/Style/Color____________

Vehicle, if any Year/Make/Model____________ Color____________

License Plate Information: State____________ Plate #____________

Instructions: Print this form and keep at least one copy with you in your vehicle, purse, or briefcase. Use this form as a general guide, completing as soon as you can following the suspicious incident you witnessed. Proceed to a safe location first – before you begin to complete this form. As soon as you have written all of the details you can recall, contact the police immediately.

- Use the back for any drawings, if applicable.
- Always keep a copy of this form for your records.

Compliments of the Northeast Intelligence Network / HQ INTEL-ALERT
www.HomelandSecurityUS.com (A Private Company)

Source: Hagman, Douglas J. 2005. Identifying potential terrorists and sleeper cells in America. North East Intelligence Network. http://www.homelandsecurityus.com/site/modules/news

Appendix C: Types of Equipment for Enhancing Physical Security

Type of Equipment	Objective	Application	Where Used
Locks	Locks are used to prevent physical access to an asset.	Locks are applied on any physical asset to be protected. Most applications require some type of strong physical structure to which the lock can be attached so that access to the asset is impeded or blocked. The use of locks also requires management of authorized access to the lock, which could include distribution and control of keys to the locks, distribution and control of combinations to the lock, management of data allowing the lock to be opened, etc.	Used on any physical asset that needs to be secured, including doors, windows, vehicles, cabinets, drawers, equipment, etc.
Alarms	Alarm systems are used to notify utility, security, or emergency personnel when a specific type of event has occurred. Events that may generate alarms can include intruders attempting to access an asset (i.e., intruder alarms); fire or other hazards (fire, smoke, or explosive vapor alarms); or other types of events.	Alarms can be applied in any number of ways. Fire and smoke alarms are required in most buildings to alert personnel to fires. Intrusion alarms are applied at any location that may be a target of unauthorized intrusion. For example, an intruder detection system (IDS) may be installed at an unmanned pump station. The IDS can be set up to send an alarm signal to a central monitoring station, allowing operators to detect unauthorized access to the pump station even though the pump station is unmanned.	Alarms are located in areas where assets need to be monitored, and thus they can be located at entrances (intrusion alarms on doors or windows), in offices or storage areas (fire, smoke, or explosive vapor alarms), in process areas (process alarms), or in any other area that needs to be protected.
Backflow prevention devices	Prevent contamination from water flowing backwards in a water distribution system.	Backflow prevention devices are typically installed at potentially hazardous cross connections to a public water system.	Finished water pipe connections and cross connections; wash basins and service sinks; and potentially hazardous, contaminant-containing systems.

Type of Equipment	Objective	Application	Where Used
Card Identification/ Access	Identify personnel as belonging to a certain group by virtue of having a special card, and then link some sort of system such that the card is required to activate the system.	Card systems have many applications. Individual cards can be used to authorize certain activities (i.e., a card entry access system); or to indicate that an individual has been in a certain place at a certain time (i.e., location tracking card systems).	Individuals must carry their cards with them and use them at required points at a facility; card readers are located at or near the assets to which they are linked.
Electronic controllers	Electronic controllers are used to automatically activate equipment (such as lights, surveillance cameras, audible alarms, or locks) when they are triggered. Triggering could be in response to a variety of scenarios, including tripping of an alarm or a motion sensor; breaking of a window or a glass door; variation in vibration sensor readings; or simply through input from a timer.	An electronic controller can be used to automate sequences of events once the controls are triggered via any of the connected input devices. The events could be all automated simultaneously or individually, depending on the controller's response to the trigger. Types of equipment that could be automated by controllers could include interior or exterior lights, facility-wide alarms, video surveillance systems, door locks, or any other equipment that can be programmed to activate based on an electronic signal.	Electronic controllers are usually located in a central location, such as in a building control room; however, they can be located anywhere there is a power supply. For enhanced system security, the control unit is usually located in an area that is secure and out of public reach so that only authorized personnel have access to it. These systems can be used to activate equipment located in any part of a system, including local and remote operations. When wireless devices are used to transmit signals to the controller, the receiver must be hardwired to the control panel.
Biometric security systems	Biometric security systems are used to control access to an asset (for example, to an entryway or a computer) by requiring that a person positively identify themselves through their unique biological characteristics before they are allowed access to that asset.	Biometrics can be applied for any system that requires the unique identification of individuals. For example, biometrics can be applied at a doorway so that only authorized individuals can gain access through that doorway, or they could be applied to a computer log-in system, so that only authorized individuals can log in to the system. While current applications are typically in high security areas, decreasing costs have resulted in systems used more frequently for lower-security applications.	The scanner component of a biometric system must be located at, on, or near the asset requiring controlled access. The central processing unit can be located either in the same location, or at a remote location.

Type of Equipment	Objective	Application	Where Used
Exterior intrusion-buried sensors	Monitor asset perimeters to detect intruders.	Designed to detect attempted physical access to a water/wastewater asset. Can be connected to an alarm, lights, or video surveillance cameras to alert facility personnel of attempted access.	Buried in ground around perimeter of asset.
Biometric hand and finger geometry recognition	Hand and finger geometry recognition security systems are used to positively identify an individual using the unique biological characteristics of their hand/fingers.	Hand and finger geometry recognition security systems can be used in any application requiring the unique verification of individuals, including time and attendance applications (i.e., tracking when an individual enters and exits a location), or access control applications. For example, hand and finger recognition can be applied at a doorway so that only authorized individuals can gain access through that doorway.	The scanner component of the hand or finger recognition security system must be located at, on, or near the asset being controlled (i.e., a secure doorway for access control applications; a log-in/log-out area for time and attendance applications). The central processing unit can be placed either in the same location, or at a remote location.
Fences	Physically deter potential intruders from gaining access to an asset.	Installed around the perimeter of any water or wastewater asset to deter unauthorized access to that asset. Fences are often placed around the perimeter or boundary of a facility, or around the perimeter of a sensitive structure within a facility. Access to the asset is controlled by directing all traffic through specific access points (e.g., gates or doors).	Perimeter of any asset to be protected.
Fence-associated exterior intrusion sensors	Monitor fence disturbances to detect intruders attempting to penetrate the fence.	Fence-associated intrusion sensors are integrated with fences to provide both a physical barrier to deter intruders and the ability to detect intruders attempting to penetrate the fence.	Fence-associated intrusion sensors are used at perimeter fence lines. They can be installed on existing fences or implemented during new fence construction. Other fence sensors can be installed as stand-alone applications in place of a traditional fence.

Type of Equipment	Objective	Application	Where Used
Films for glass shatter protection	Protect windows, glass doors, and other glass from shattering.	Can be used on any glass surface to prevent the glass from shattering. Preventing the glass from shattering may prevent access to a building or a room through the broken glass, and may also help to reduce injuries to personnel located behind the glass.	Windows, glass doors, and any other piece of glass at a water/wastewater utility.
Fire hydrant locks	Hydrant locks can be used to prevent unauthorized physical access to a water asset via a fire hydrant.	Installing hydrant locks on all hydrants in a system may help to prevent introduction of unauthorized substances into the system.	On any fire hydrant valve.
Ladder access control	To delay access to assets such as roofs, raised water tanks, pipes, or other assets by controlling access to the ladders leading to the asset.	Used to protect any indoor or outdoor ladder.	On any indoor or outdoor ladder, such as ladders leading to roofs, water tanks, raised pipes, or other raised assets.
Manhole intrusion sensors	To detect unauthorized intrusion into a manhole through the use of sensors located in the manhole.	Installing intrusion sensors on all manholes in a system may help to prevent unauthorized personnel from accessing or entering the system. Monitoring manholes may also prevent the introduction of hazardous substances into the storm water or wastewater system.	The sensors are located within the manhole; alarms are communicated to a central monitoring location.
Manhole locks	Manhole locks can be used to prevent unauthorized physical access to sewer lines, water valves, or other water or wastewater assets via a manhole.	Installing manhole locks on all manholes in a system may help to prevent unauthorized personnel from accessing or entering the system. Locking manholes may also prevent the introduction of hazardous substances into the storm water or wastewater system.	On any type of manhole.
Reservoir covers	Protect water supplies from unauthorized physical access by installing a cover over the water surface. Covering the reservoir may also reduce the potential for accidental introduction of contaminants into the reservoir.	Covers can be applied to any existing open-air water reservoir, depending on its size and structural characteristics.	The cover is physically placed over all or part of the reservoir to ensure that the water supply is not accessible from the surface.

Type of Equipment	Objective	Application	Where Used
Security barriers	To prevent vehicles from accessing an area.	Security barriers can be placed anywhere along roadways or in front of buildings. For example, a security barrier can keep a car from crashing into a building, or keep intruders from entering a facility in a car.	Security barriers are used in or along roads, sidewalks, paths, or perimeters.
Security for doorways side hinged doors	Protect a door from being forcefully entered. Security of the doorway can be enhanced by modifying the door, the doorframe, the hinges, or the lock. Different doorway security measures may protect against various potential threats, including breaking, blasting, or fire.	Can be applied to any doorway. An individual application may consist of modifying one or more features of the doorway, depending on the potential threats.	Used in any doorway that may be a target for intruders.
Valve lockout devices	To prevent or delay unauthorized access to a valve.	Used to ensure a valve remains in the desired position and is not tampered with by an unauthorized individual. Valve lockout devices are placed on, over, or through valve handles to prevent rotation.	On valves located in water or wastewater treatment plants, remote facilities (pumping stations, etc.), water distribution systems, backflow prevention devices, and other piping systems.
Visual surveillance monitoring	Visually monitor an asset to detect potential intruders, unauthorized or suspicious materials or objects, or other threats.	Used to detect physical threats to an asset (i.e., persons or materials) through surveillance of asset. Can be used to monitor any water or wastewater assets (perimeter of facility, remote pump houses, potential access points to distribution or collection systems, etc.). Primarily used to monitor exterior areas, but can be used in interior of buildings or facilities.	Usually mounted at a strategic location at the asset to be monitored to monitor as large an area as possible. Can be mounted near doors or windows, on or along fences, or within buildings.

Source: EPA. Water Security Products Guide. http://cfpub.epa.gov/safewater/watersecurity/guide

Index

C

F

G

H

M

N

R

S

T

U

V

X

Y

Z